Selected Titles in This Series

D0986028

Volume

11 **Conference Board of the Mathematical Sciences**
The mathematical education of teachers
2001

10 **Solomon Friedberg et al.**
Teaching mathematics in colleges and universities: Case studies for today's
classroom. Available in student and faculty editions
2001

9 **Robert Reys and Jeremy Kilpatrick, Editors**
One field, many paths: U. S. doctoral programs in mathematics education
2001

8 **Ed Dubinsky, Alan H. Schoenfeld, and Jim Kaput, Editors**
Research in collegiate mathematics education. IV
2001

7 **Alan H. Schoenfeld, Jim Kaput, and Ed Dubinsky, Editors**
Research in collegiate mathematics education. III
1998

6 **Jim Kaput, Alan H. Schoenfeld, and Ed Dubinsky, Editors**
Research in collegiate mathematics education. II
1996

5 **Naomi D. Fisher, Harvey B. Keynes, and Philip D. Wagreich, Editors**
Changing the culture: Mathematics education in the research community
1995

4 **Ed Dubinsky, Alan H. Schocnfeld, and Jim Kaput, Editors**
Research in collegiate mathematics education. I
1994

3 **Naomi D. Fisher, Harvey B. Keynes, and Philip D. Wagreich, Editors**
Mathematicians and education reform 1990–1991
1993

2 **Naomi D. Fisher, Harvey B. Keynes, and Philip D. Wagreich, Editors**
Mathematicians and education reform 1989–1990
1991

1 **Naomi D. Fisher, Harvey B. Keynes, and Philip D. Wagreich, Editors**
Mathematicians and education reform: Proceedings of the July 6–8, 1988 workshop
1990

The Mathematical
Education of Teachers

CBMS

Conference Board of the Mathematical Sciences

Issues in Mathematics Education

Volume 11

The Mathematical
Education of Teachers

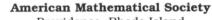

 American Mathematical Society
Providence, Rhode Island
in cooperation with
Mathematical Association of America
Washington, D. C.

United States Department of Education grant number R309U97001 made to the Mathematical Association of America in support of "The Mathematics Education of Teachers Project" is gratefully acknowledged.

2000 *Mathematics Subject Classification.* Primary 97–XX.

Library of Congress Cataloging-in-Publication Data

The mathematical education of teachers / Conference Board of the Mathematical Sciences.
    p. cm. — (Issues in mathematics education, ISSN 1047-398X ; v. 11)
    Includes bibliographical references.
    ISBN 0-8218-2899-1 (alk. paper)
    1. Mathematics—Study and teaching (Higher)—United States. 2. Mathematics teachers—Training of—United States. I. Conference Board of the Mathematical Sciences. II. Series.

QA13.M1515   2001
510′.7′1073—dc21
                                                      2001045098

Steering Committee

James Lewis (chair)

Richelle Blair

Gail Burrill

Joan Ferrini-Mundy (advisor)

Roger Howe

Mary Lindquist

Carolyn Mahoney

Dale Oliver

Ronald Rosier (ex-officio)

Richard Scheaffer

Writing Team

Alan Tucker (lead writer)

James Fey

Deborah Schifter

Judith Sowder

Editing Team

Cathy Kessel (lead editor)

Judith Epstein

Michael Keynes

Table of Contents

Preface   xi

PART 1

Chapter 1. Changing Expectations, New Realizations   3

Chapter 2. General Recommendations   7

Introduction to Recommendations for Teacher Preparation   13

Chapter 3. Recommendations for Elementary Teacher Preparation   15

Chapter 4. Recommendations for Middle Grades Teacher Preparation   25

Chapter 5. Recommendations for High School Teacher Preparation   37

Chapter 6. Recommendations for Technology in Teacher Preparation   47

References   49

Appendix: Relevant Reports   51

PART 2

Chapter 7. The Preparation of Elementary Teachers   55

Chapter 8. The Preparation of Middle Grades Teachers   99

Chapter 9. The Preparation of High School Teachers   121

Preface

This report is designed to be a resource for mathematics faculty and other parties involved in the education of mathematics teachers. It is a distillation of current thinking on curriculum and policy issues affecting the mathematical education of teachers, with the goal of stimulating efforts on individual campuses to improve programs for prospective teachers. It is also intended to marshal the backing of the mathematical sciences community for important national initiatives, such as the use of mathematics specialists to teach mathematics starting in middle grades and expanded time for professional development in the schools.

Now is a time of great interest in K–12 mathematics education. Student performance, curriculum, and teacher education are the subjects of much scrutiny and debate. Studies of the mathematical knowledge of prospective and practicing U. S. teachers suggest ways to improve their mathematical educations.

Two general themes of this report are: (i) the intellectual substance in school mathematics; and (ii) the special nature of the mathematical knowledge needed for teaching. It is often assumed that, because the topics covered in school mathematics are so basic, they must also be easy to teach. We owe to mathematics education research of the past decade, or so, the realization that substantial mathematical understanding is needed even to teach whole number arithmetic well. Several mathematics education researchers, in particular Deborah Ball and Liping Ma, have been able to communicate these findings in ways that engaged research mathematicians. Middle grades curricula are even more demanding; for example, the structure of the rational numbers and the idea of proportionality require even more knowledge of teachers. High school mathematics is often considered more substantive than the mathematics of earlier grades, but the challenges of developing a knowledge of it for teaching are often unacknowledged.

The mathematical knowledge needed for teaching is quite different from that required by college students pursuing other mathematics-related professions. Prospective teachers need a solid understanding of mathematics so that they can teach it as a coherent, reasoned activity and communicate its elegance and power. Mathematicians are particularly qualified to teach mathematics in the connected, sense-making way that teachers need. For maximum effectiveness, the design of this instruction requires collaboration between mathematicians and mathematics educators and close connections with classroom practice.

This report is not aligned with a particular school mathematics curriculum, although it is consistent with the National Council of Teachers of Mathematics' *Principles and Standards for School Mathematics* as well as other recent national reports on school mathematics. This report focuses on preservice education. Although there is a growing awareness that teachers need more professional development opportunities, this project did not have the time or resources to discuss this important issue in detail.

This report is addressed to a number of different audiences. The following paragraphs offer guidance to readers in some of these audiences.

Mathematics faculty. The primary audience for this report are the members of the mathematics faculties of two-year colleges, four-year colleges, and universities. In this report, the terms "mathematicians" and "mathematics faculty" refer to all mathematical scientists, including statisticians. Teacher education is an important component of the educational mission of most mathematics departments. Many faculty members of these departments, not just the teacher education specialists, ought to have an appreciation of the curriculum and policy issues that affect prospective and practicing teachers. Part 1 of this report, Chapters 1–6, attempts to give mathematics faculty a useful summary of these issues. In departments where some mathematics majors are prospective high school mathematics teachers, mathematicians who teach courses in the major are also encouraged to read Chapter 9 for information about the needs of prospective teachers in their courses.

Mathematics education faculty and mathematics faculty deeply involved in teacher education. Part 2 of this report, Chapters 7–9, is aimed foremost at these faculty. These chapters go into considerable detail about major themes in these courses, including strategies for developing a sound mathematical understanding and an awareness of common student misconceptions. On the other hand, these chapters do not spell out complete topic-by-topic syllabi for mathematics courses for teachers. The creation of revised and new courses for teachers should be the product of extensive local and national discussion and experimentation, built on the ideas in these chapters. Considerable variation from institution to institution is likely.

Mathematics chairs and deans. The general recommendations in Chapter 2 are the most important part of the report for these readers. Mathematics chairs also need to be familiar with the themes of Chapters 3–5, which summarize recommendations about the mathematics courses for prospective teachers at different grade levels.

Mathematics supervisors in schools, district offices, and state education departments. Although the primary audience for this report is mathematics faculty, it is important that the mathematical education of teachers be recognized both inside and outside of colleges and universities as a collaboration involving all interested parties. Parts 1 and 2 of this report are meant to build a foundation for such collaboration.

Education policy bodies at the state and national level. Policy issues are summarized in Chapter 2. This chapter is meant to serve as a two-way street between mathematicians and state and national education policy bodies. It presents the policy concerns of mathematicians, which need support from policy bodies, and recommendations from these bodies, which mathematicians are encouraged to support.

Organizations of mathematicians. Professional organizations in the mathematical sciences have a critical role to play in the mathematical education of teachers, by fostering discussion and encouraging greater involvement among their members. They can organize workshops on this topic at professional meetings and maintain Web sites with information on teacher education and related issues.

Organizations that fund efforts to improve teacher education. This report asks mathematicians to rethink courses on fundamental ideas of school mathematics, and in particular, that they give more attention to the special nature of the mathematical knowledge needed for teaching. It is imperative that federal agencies and private foundations fund a wide array of efforts to develop successful models of such courses and support independent assessments of such models. Support for workshops to disseminate their results is also necessary.

Accreditation and certification organizations. Agencies that certify teachers and accredit teacher education programs need to work in careful coordination with mathematics and mathematics education faculties, as the rethinking suggested here evolves. This report does not offer detailed standards for teacher knowledge of mathematics that can be readily translated into certification requirements. However, it does provide considerable guidance about the kind of mathematical understanding that teachers need and the connection of that understanding with day-to-day classroom instruction.

The writers of this report hope that it will be useful to other audiences—both current and future teachers, education faculty in general, and school supervisors of mathematics.

This document is available in print and electronic formats. Part 1 is published by the Mathematical Association of America and may be obtained free of charge from the Conference Board of the Mathematical Sciences. The American Mathematical Society has published Parts 1 and 2 as a single volume. Parts 1 and 2 may also be downloaded from the Conference Board of the Mathematical Sciences Web site, `http://www.maa.org/cbms`.

Acknowledgements. The official writers of this report are listed at the front of this document, along with the members of the project steering committee. However, the focus, language, and overall tone of this report reflect the helpful advice of scores of mathematicians and mathematics educators, who commented on the many drafts that have circulated over the past two years. Although the writers sometimes had to choose among conflicting advice, it was heartening to see how much consistency and agreement there was in most of the comments.

PART 1

Chapter 1

Changing Expectations, New Realizations

This report calls for a rethinking of the mathematical education of prospective teachers within mathematical sciences departments at U. S. two- and four-year colleges and universities. It offers principles to assist departments in this process, along with specific suggestions for mathematics courses for prospective teachers. Additionally, this report seeks to convince faculty that there is more intellectual content in school mathematics instruction than most realize, content that teachers need to understand well.

Opinion polls show that education currently is the number one concern of American adults. A 1998 Harris Poll revealed that roughly nine out of ten Americans believe that the best way to raise student achievement is to ensure a qualified teacher in every classroom. States are mandating higher standards for teacher certification, as well as for student achievement, and they are monitoring teacher preparation programs more closely. School mathematics instruction and the mathematical preparation of teachers are in the spotlight, because, after reading and writing, mathematics is widely viewed as the most important component of K–12 education to promote future success in college and subsequent careers. All mathematicians[1] should be concerned about teacher education, and all have a role to play in setting policies, if not offering instruction, in the mathematical education of teachers.

It goes without saying that mathematicians want prospective teachers to have a solid understanding of the mathematics that they will teach. The daunting challenge is how to achieve this goal, given the diverse mathematical preparation of college students who will become teachers and the changing views about what mathematical knowledge is needed to be an effective teacher. Mathematicians who teach introductory mathematics courses commonly encounter too many students who are afraid of mathematics, lack needed study skills, or have deficiencies in their mathematical background. On top of these problems, mathematicians who teach courses for prospective teachers face the additional challenge of making appropriate connections to mathematical instruction in school classrooms.

A number of mathematicians and mathematics education researchers have recognized the special nature of the mathematical knowledge needed for K–12 teaching and its implications for the mathematical preparation of teachers. In particular,

---

[1]For simplicity of language, this report will use the terms "mathematics" and "mathematicians" to refer to the mathematical sciences, including statistics, and to mathematical scientists, respectively. Likewise, "mathematicians" refers to faculty at two-year colleges, four-year colleges, and universities.

the interviews with Chinese elementary teachers in Liping Ma's 1999 book *Knowing and Teaching Elementary Mathematics* awakened many mathematicians to this issue and its mathematical substance. For example, mathematical knowledge for teaching should prepare an elementary teacher to assess the validity of the following subtraction procedure proposed by a student:

$$43 - 27 = (40 - 20) \text{ (tens place subtraction)} + (3 - 7) \text{ (ones place subtraction)}$$
$$= \quad 20 \qquad\qquad\qquad\qquad + \quad -4$$
$$= \quad 16$$

and to appreciate the complications of trying to extend it to larger numbers.

Some aspects of mathematical knowledge for teaching, such as those illustrated in the example above, may seem to mathematicians to fall into the domain of methods courses in education. However, education faculty generally see these issues to be more appropriately addressed in mathematics courses, and so such issues often remain unaddressed in teacher preparation. This state of affairs is one of many reasons why efforts to improve the mathematical education of teachers require a partnership between faculty in mathematics and mathematics education.

Changing expectations for mathematics knowledge. Arithmetic skills, and occasionally a little algebra, were once the mathematics required for almost all jobs outside of engineering and the physical sciences. In recent years, computers and an associated explosion in the use of quantitative methods in business and science have dramatically increased the mathematical skills needed in many jobs. Facility at creating spreadsheets is becoming required in many entry-level positions for high school graduates. Assembly line workers may be expected to learn elements of statistical quality control. The level of mathematical sophistication common in financial analyses today would have been unthinkable a generation ago.

Changing expectations for school mathematics instruction. Public education in the United States has historically had a utilitarian focus, which in mathematics, emphasized arithmetic skills and problems from commerce, such as compound interest. Until recently, only high school students in college preparatory tracks studied algebra, and then often just for one year. In the two decades after World War II, there were efforts to increase the sophistication of mathematics curricula in colleges and schools. These efforts succeeded in moving calculus into the first year of college and modernizing the curriculum of mathematics majors. The school mathematics reform efforts of the 1960s and 1970s had a number of long-lasting influences, such as broadening elementary school mathematics beyond arithmetic to include some geometry and elements of algebra, and refocusing high school mathematics by downplaying analytic geometry and trigonometry and giving more attention to functions and providing an introduction to calculus. However, this period is most remembered for the New Math movement's theoretical approach. This approach was widely rejected, leaving school mathematics reform efforts on the defensive for many years to come.

There have been complaints about the poor mathematical skills of students throughout the history of U. S. education. An unacceptably low number of students today are making their way up the mathematical ladder to the higher levels of mathematical knowledge that their college majors and future employment require. There are growing concerns about what students are being taught, as well as how well they are learning it. Since the 1980s, international comparison studies, such as the Third International Mathematics and Science Study, have indicated that many developed countries, particularly some in east Asia, provide school mathematics instruction richer than that of the United States (see, e.g., Stevenson and Stigler's book *The Learning Gap* and McKnight et al.'s *The Under-achieving Curriculum*).

In 1989, the National Council of Teachers of Mathematics (NCTM) *Curriculum and Evaluation Standards* initiated an overdue period of intense interest in strengthening school mathematics instruction, to make it both more demanding and more effective for all students. Given the immense problems in both primary and secondary schools which affect both the development of procedural and reasoning skills, it was probably inevitable that there would not be a consensus on the precise nature of some shortcomings in current mathematics instruction, nor on proposed solutions. In 2000, the NCTM *Principles and Standards for School Mathematics*, an update of the 1989 *Standards*, addressed the issues of interpretation and implementation that had emerged over the preceding decade, including input from a broad spectrum of mathematical science and mathematics education organizations.

New realizations about the mathematical education of teachers. Throughout U. S. educational history, teachers have generally provided the style and level of instruction that society expected of them. Until 1900, teachers of mathematics were largely seen as drill masters, training students to accurately perform numerical computations. Beyond the eight primary grades, most teachers had at best a year or two of preparation at a special high school, called a normal school. The introduction of universal high school around 1900 gave rise to secondary level subject specialists, who majored in their subject in teachers' colleges. Teachers for earlier grades also were eventually required to go to college, but their education focused on the psychological and social development of children. It was generally assumed, and is still assumed by some today, that prospective elementary school teachers, and perhaps middle school teachers, learn all the mathematics they need to teach mathematics well during their own schooling.

Recently, this assumption has been seriously questioned. There is evidence of a vicious cycle in which too many prospective teachers enter college with insufficient understanding of school mathematics, have little college instruction focused on the mathematics they will teach, and then enter their classrooms inadequately prepared to teach mathematics to the following generations of students. Studies of teachers' mathematical knowledge, for example Ball's 1991 "Research on teaching mathematics," have drawn attention to substantial mathematical issues that arise in day-to-day school instruction, but are not well understood by prospective U. S. teachers, when they graduate from college. One example is the place value structure of our number system, which implicitly expresses numbers as polynomials in powers of 10 and permits single-digit arithmetic to be easily extended to multi-digit arithmetic (in contrast with Roman numeral computations). Elementary teachers need a deep understanding of place value to help their students use it as a foundation for the successful learning of integer arithmetic, and later decimal arithmetic and symbolic calculations in algebra. Likewise, middle grades teachers need a deep understanding of proportions to help their students master fractions, and high school teachers need a deep understanding of functions to help prepare their students for the future study of calculus.

International studies have also highlighted the importance of continuing study as an integral part of a teacher's weekly duties. Thus, college mathematics courses should be designed to prepare prospective teachers for the lifelong learning of mathematics, rather than to teach them all they will need to know in order to teach mathematics well.

In sum, these new expectations and realizations make a strong case for a thorough rethinking of mathematics courses for prospective teachers of all grade levels.

Chapter 2

General Recommendations

This chapter presents the core recommendations to mathematics departments and the general mathematics community about the mathematical education of teachers. These recommendations fall into three categories: curriculum and instruction for prospective teachers in mathematics departments, cooperation between mathematics departments and other parties involved with the education of teachers, and mathematicians' involvement in national policy that supports high quality mathematics teaching.

Mathematics Curriculum and Instruction for Prospective Teachers

Recommendation 1. Prospective teachers need mathematics courses that develop a deep understanding of the mathematics they will teach. The mathematical knowledge needed by teachers at all levels is substantial, yet quite different from that required by students pursuing other mathematics-related professions. Prospective teachers need to understand the fundamental principles that underlie school mathematics, so that they can teach it to diverse groups of students as a coherent, reasoned activity and communicate an appreciation of the elegance and power of the subject. With such knowledge, they can foster an enthusiasm for mathematics and a deeper understanding among their students. College courses developing this knowledge should make connections between the mathematics being studied and mathematics prospective teachers will teach. Finally, prospective teachers need to develop a thorough mastery of the mathematics in several grades beyond that which they expect to teach, as well as of the mathematics in earlier grades.

Courses about the fundamental ideas of school mathematics should be taught by mathematicians who have a serious interest in teacher education. This instruction should be coordinated with faculty in mathematics education. It is vital that faculty in the mathematics department and the school of education agree on substantial expectations for student learning and achievement in these mathematics courses. Further, careful thought needs to be given to appropriate prerequisite knowledge for these courses. Special courses for teachers, rather than traditional courses, such as intermediate algebra and college algebra, are desirable for those lacking prerequisites.

Recommendation 2. Although the quality of mathematical preparation is more important than the quantity, the following amount of mathematics coursework for prospective teachers is recommended.

(i) *Prospective elementary grade teachers should be required to take at least 9 semester-hours on fundamental ideas of elementary school mathematics.*

(ii) *Prospective middle grades teachers of mathematics should be required to take at least 21 semester-hours of mathematics, that includes at least 12 semester-hours on fundamental ideas of school mathematics appropriate for middle grades teachers.*

(iii) *Prospective high school teachers of mathematics should be required to complete the equivalent of an undergraduate major in mathematics, that includes a 6-hour capstone course connecting their college mathematics courses with high school mathematics.*

Chapters 3–5 give recommendations for mathematics courses for prospective elementary, middle, and high school teachers. Chapters 7–9 (Part 2 of this report) develop these recommendations in greater detail.

Recommendation 3. Courses on fundamental ideas of school mathematics should focus on a thorough development of basic mathematical ideas. All courses designed for prospective teachers should develop careful reasoning and mathematical "common sense" in analyzing conceptual relationships and in solving problems. Attention to the broad and flexible applicability of basic ideas and modes of reasoning is preferable to superficial coverage of many topics. Prospective teachers should learn mathematics in a coherent fashion that emphasizes the interconnections among theory, procedures, and applications. They should learn how basic mathematical ideas combine to form the framework on which specific mathematics lessons are built. For example, the ideas of number, geometry, and function, along with algebraic and graphical representation of information, form the basis of most high school algebra and trigonometry.

Recommendation 4. Along with building mathematical knowledge, mathematics courses for prospective teachers should develop the habits of mind of a mathematical thinker and demonstrate flexible, interactive styles of teaching. Mathematics is not only about numbers and shapes, but also about patterns of all types. In searching for patterns, mathematical thinkers look for attributes like linearity, periodicity, continuity, randomness, and symmetry. They take actions like representing, experimenting, modeling, classifying, visualizing, computing, and proving. Teachers need to learn to ask good mathematical questions, as well as find solutions, and to look at problems from multiple points of view. Most of all, prospective teachers need to learn how to learn mathematics.

Results of international studies, as described, for example, in Stigler and Hiebert's 1999 book *The Teaching Gap*, indicate that U. S. school mathematics instruction places a comparatively low priority on engaging students to develop an understanding of mathematics. To foster more of this activity in schools, prospective teachers need to experience such instruction in their college mathematics classes and to learn that there are multiple ways to engage students in mathematics.

Recommendation 5. Teacher education must be recognized as an important part of mathematics departments' mission at institutions that educate teachers. More mathematicians should consider becoming deeply involved in K–12 mathematics education. Mathematics departments should devote commensurate resources to designing and offering courses for teachers. They should also value and properly reward the faculty members heavily involved in teacher education. Whether or not a mathematics department contains a group of mathematics education specialists, it is important for the entire mathematics faculty to actively support teacher education efforts. In return, mathematics departments should receive the resources needed to follow through on their commitment to high quality teacher education. As argued in the 1999 American Mathematical Society report *Towards Excellence*, a strong commitment to issues of societal concern, such as teacher education, can help garner administrative support for other departmental priorities.

Senior mathematics faculty, who have become involved in K–12 mathematics education, should see themselves as possible models for fellow faculty, and their efforts should be publicized. All mathematics faculty should value, encourage, and support interest among students in pursuing a career in teaching.

Cooperation Among Parties Involved in Teacher Education

Recommendation 6. The mathematical education of teachers should be seen as a partnership between mathematics faculty and mathematics education faculty. Most good school mathematics instruction involves a combination of mathematical knowledge and pedagogy, such as choosing appropriate examples and teaching strategies for various topics. Mathematics educators can provide valuable insights and information about what takes place in school classrooms, including common mathematical misunderstandings of practicing teachers and how to build on these to promote real understanding. They have access to information on state curriculum guidelines and research studies about teachers' mathematical knowledge. In return, mathematics faculty can help mathematics education faculty by keeping them informed of mathematical developments which have an impact on school mathematics.

Ph.D.-granting mathematics departments are encouraged to work with mathematics education faculty to develop new Ph.D. programs in mathematics education as a response to the many unfilled faculty openings for mathematics education specialists. These two groups are also encouraged to develop "minors" in mathematics education for mathematics Ph.D.'s, similar to the statistics and computer science concentrations that some mathematics Ph.D. programs offer.

The reality today is that there is considerable distrust between mathematics faculty and mathematics education faculty both within institutions and through public exchange. Conscious efforts, locally and nationally, are needed to foster cooperation, along with mutual understanding and respect between these two groups. Mathematicians and mathematics educators, when working cooperatively, can be more effective in influencing the state and national organizations responsible for curriculum standards and certification of teacher education programs.

Recommendation 7. There needs to be greater cooperation between two-year and four-year colleges in the mathematical education of teachers. Two-year colleges are essential partners in the mathematical education of teachers. A large number of future teachers begin their post-secondary study in two-year colleges. In particular, many elementary teachers take a significant portion of their college mathematics courses at two-year colleges. Many of the new courses on school mathematics, mentioned in Recommendation 2 and discussed in subsequent chapters, should be offered in two-year colleges.

Moreover, there are special difficulties that two-year colleges face in helping to prepare teachers. Students of two-year colleges transfer to a variety of different four-year institutions with differing course requirements for prospective teachers. Good articulation agreements are needed among two-year colleges, university mathematics departments, and education departments about the mathematical coursework for prospective teachers. The 1999 National Science Foundation report *Investing in Tomorrow's Teachers* contains a comprehensive set of recommendations about the role of two-year colleges in the mathematical education of teachers.

Recommendation 8. There needs to be more collaboration between mathematics faculty and school mathematics teachers. Observing teachers in action and learning about their experiences can give mathematicians helpful perspectives for their instruction of prospective teachers. Involving school teachers in courses about school mathematics is a tangible way to connect the courses with real practice. In turn, mathematicians should have an important role in professional development activities for teachers.

Policies to Support High Quality School Mathematics Teaching

Recommendation 9. Efforts to improve standards for school mathematics instruction, as well as for teacher preparation accreditation and teacher certification, will be strengthened by the full-fledged participation of the academic mathematics community. While it is generally true that few academic mathematicians gave much attention to standards for school mathematics and teacher preparation in the 1970s and 1980s, they have historically played a major role in these arenas, and in recent years, a number have become involved again. Published standards are likely to carry more weight and generate less controversy when mathematicians, along with mathematics education researchers, curriculum developers, teachers, and other interested parties, play a significant role in their drafting.

In terms of the agenda of this report, it is hard for mathematicians to develop good programs in the mathematical education of teachers, if they are not involved in establishing standards for school mathematics instruction, accreditation of teacher preparation programs, and teacher certification.

Recommendation 10. Teachers need the opportunity to develop their understanding of mathematics and its teaching throughout their careers, through both self-directed and collegial study, and through formal coursework. In some countries where student achievement is high, teachers, alone and in groups, spend time refining their lessons and studying the underlying mathematics. They observe each other's classes. Beginning teachers have extensive mentoring. The teachers' manuals accompanying their textbooks have extensive background material about the mathematics being taught and how it fits into the overall curriculum. More professional development opportunities of this kind are essential for U. S. teachers.

Through continuing education courses, outreach to schools, and other efforts, mathematics departments should support a culture in the teaching profession that promotes professional growth. Mathematicians should speak out in support of changes in the schools to make professional development an integral part of a teacher's job.

A more complete discussion of the role of mathematics faculty in professional development concerns is left to a future study.

Recommendation 11. Mathematics in middle grades (grades 5–8) should be taught by mathematics specialists. This recommendation mirrors similar recommendations by a number of other groups seeking to improve U. S. school mathematics instruction. Middle grades mathematics teachers must know the high school mathematics curriculum well and understand the foundation that is being laid for it in their instruction. As concepts like fractions and decimals enter the curriculum, teaching mathematics well requires subject matter expertise that non-specialists cannot be expected to master. Having mathematical specialists, beginning in middle grades, both reduces the educational burden for those teaching mathematics in these grades and provides opportunities for prospective teachers of these grades who like mathematics to specialize in it.

Introduction to Recommendations for Teacher Preparation

The following three chapters on the mathematical preparation for teachers at various grade levels take a different approach than the recommendations of the 1991 MAA teacher preparation report *A Call for Change*. That report was built around a broad inventory, stated in general terms, of the mathematical knowledge and reasoning that K–12 teachers need in order to teach mathematics well. This report augments those recommendations, by giving more attention to the mathematical conceptions of K–12 students and how their teachers can be better prepared to address these ideas. The conceptions that prospective teachers often bring to college classes also get considerable attention.

Grade level. The next three chapters are organized around three common U. S. grade groupings: elementary grades (1–4), middle grades (5–8), and high school (9–12). A distinctive teacher preparation program is proposed for each grouping.

Teachers need to study the mathematics of a cluster of grade levels, both to be ready for the various ways in which grades are grouped into elementary, middle, and high schools in different school districts, and to understand the larger mathematical learning context in which the mathematics taught in a specific grade fits. Consequently, mathematics programs for teachers need some breadth in the grade levels they target. This report calls for mathematics specialists beginning at least by 5th grade, so that, for example, a mathematics specialist might teach all of the 4th and 5th grade students in a small K–5 elementary school.

Curriculum. The mathematical education of teachers proposed in the following chapters is meant to prepare knowledgeable, flexible teachers who are able to effectively educate students in mathematics, using a variety of current and future mathematics curricula. This document is particularly supportive of the NCTM *Principles and Standards for School Mathematics*, as well as other recent national reports on school mathematics.

Developing deep understanding. There are a number of statements in this report about prospective teachers acquiring a "deep understanding" of school mathematics concepts and procedures. The emphasis is on the mathematics that teachers need to know but also there is a recognition that teachers must develop "mathematical knowledge for teaching." This knowledge allows teachers to assess their students' work, recognizing both the sources of student errors and their students' understanding of the mathematics being taught. They also can appreciate and nurture the creative suggestions of talented students. Additionally, these teachers see the links between different mathematical topics and make their students aware of them. Such teachers are also more able to excite students about mathematics. Some mathematicians may react skeptically to setting these goals for prospective teachers, because, in their experience, prospective teachers, like many other students in introductory mathematics courses, seem to struggle to gain a minimal understanding of the basic concepts. Indeed, it is only realistic to expect such knowledge to develop over years of professional study, undertaken alone, with other teachers, and in continuing education classes. However, its foundation—deep understanding of school mathematics—must be laid during preservice education.

Chapters 3, 4, and 5 in Part 1 of this report outline this mathematical foundation. Chapters 7, 8, and 9 (Part 2 of this report) describe this mathematical knowledge in greater detail and illustrate how the need for it arises in teaching. Part 2 is intended for those faculty members who will teach courses in the foundations of school mathematics or who want to broaden their backgrounds in school mathematics instruction. These chapters are also intended to be useful to mathematicians, in general, to make them aware of the pedagogical issues connected with "mathematical knowledge for teaching."

Proof and justification. Mathematicians need to help prospective teachers develop an understanding of the role of proof in mathematics. In the Reasoning and Proof standard, *Principles and Standards for School Mathematics* says "Proof is a very difficult area for undergraduate mathematics students. Perhaps students at the postsecondary level find proof so difficult because their only experience in writing proofs has been in a high school geometry course." Prospective teachers at all levels need experience justifying conjectures with informal, but valid arguments if they are to make mathematical reasoning and proof a part of their teaching. Future high school teachers must develop a sound understanding of what it means to write a formal proof.

Terminology. To avoid confusion, the report uses the following terminology:

Student refers to a child in a K–12 classroom.

Teacher refers to an instructor in a K–12 classroom, but may also refer to a prospective K–12 teacher in a college mathematics course ("prospective teacher" is also used in the latter case).

Instructor refers to an instructor of prospective teachers. In this report, that person will usually be a mathematician.

Chapter 3

Recommendations for Elementary Teacher Preparation

Is elementary mathematics so simple that teaching it requires knowing only the "math facts" and a handful of algorithms? The premise of this chapter and its elaboration in Part 2 is that, quite to the contrary, this early content is rich in important ideas. It is during their elementary years that young children begin to lay down those habits of reasoning upon which later achievement in mathematics will crucially depend. Thus, for example, it is unrealistic to expect students who failed to develop early an understanding of how to manipulate arithmetic expressions to later manipulate algebraic expressions with confidence. And those students who have never had experience with decomposing and recomposing shapes in their early education are unlikely to attach meaning to the succession of assertions in typical proofs in Euclidean geometry.

When the goal of instruction is to help children attain both computational proficiency and conceptual understanding, teaching elementary school mathematics can be intellectually challenging. Consider the following vignette from a third grade classroom:[1]

> The children have been working on multiplication, exploring what the operation *means*—the kinds of situations it models—in addition to learning their multiplication facts. Now, as they approach multiplication of two-digit numbers, their teacher wants to identify the ideas they bring to this new topic. She gives her students a problem— *There were 64 teams at the beginning of the NCAA basketball tournament. With 5 players starting on each team, how many starting players were in the tournament?*—and her students offer a variety of solution methods:

> Laurel: That would be 64×5. I use one 10 because I know $5 \times 10 = 50$. Then you do that six times. That's 30, I mean 300. Then you add 4 five times, which is 25, no 20. I added it all together and got 320.

> Chris: 64 means $60 + 4$. So I did 60 five times, for 300. Then 4×5 is 20, so the answer is 320.

> Jack: I split 64 into four parts—[First, I did] 20, 20, and 20. I did each one separately: $20 \times 5 = 100$, $20 \times 5 = 100$, $20 \times 5 = 100$. Then the last part, 4×5, is 20. All together, 320.

---

[1]This vignette has been drawn from an actual event. More detail is given in Schifter et al., 1999, pp. 82-86.

The teacher now begins to distribute manipulatives for the next activity, but decides, on the spur of the moment, to give a new problem: *We have 18 kids here today and each needs 12 tiles for the next activity. How can we figure out the number of tiles to give out?*

Josh: That would be 18 × 12, and I know 10 × 10 is 100 and 8 × 2 is 16, so if you add them together it would be 100 + 16 = 116.

Dava: That's wrong. I did 18 × 10 and got 180, but I thought at first I was wrong, so I double checked. I noticed that Josh didn't do 8 × 10, so my answer [for the sub-product, 18 × 10] was right. I didn't do the 2 yet, so I do 18 × 2. Then you add it up—180 + 36.

Presented with this scene from an elementary classroom, some readers may wonder why the teacher has solicited these children's ideas about how to multiply two-digit numbers before showing them the standard procedure. All the teacher really needs to do is take her students through the algorithm step by step—didn't we learn it that way? Relinquish, for a moment, memories of your own elementary-school experience and consider the opportunities for learning this classroom offers the children.

Implicit in the methods proposed by Laurel, Jack, Chris, and Dava, even Josh, are the associative and distributive properties and recognition of the flexibility gained by decomposing numbers into tens and ones. Thus, what appears to have been the initial step of Josh's strategy for multiplying 18 × 12—think of 18 as 10 + 8, 12 as 10 + 2—is sound reasoning. However, perhaps relying too exclusively on his understanding of additive relationships, his answer fails to account for all of the necessary sub-products. He has fallen victim to a mistake similar to one so common in college calculus classes that it has its own name, "the Freshman's Dream": the belief that $(a + b)(c + d) = ac + bd$.

Now where is the teacher to go with all these ideas: compare the strategies of the children who got things right, explore Josh's procedure to see where he went wrong, or continue on to the next planned activity, perhaps later to come back to Josh and Dava? What should go into making such a decision, what must their teacher understand in order to work successfully with these children's ideas?

First, she must believe that mathematics is about ideas that make sense, rather than a collection of motiveless rules, and that her students have mathematical ideas that can be built upon; next, that there are many ways to solve a given problem. Then, she must be able to follow her students' thinking to determine which of the solution methods they propose are valid and identify the concepts upon which those methods are built. Too, she must recognize not only that Josh has made an error, but be able to subject the reasoning behind his error to investigation in a variety of ways. For example, what if 18 times 12 tiles were actually doled out? And how would this result compare to what an area representation of 18 × 12 would show?

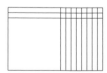

An area representation of 18×12

More abstractly, how does the source of Josh's error connect to additive procedures, the distributive property, multiplication algorithms, multiplication of binomials, and so on? And how can these connections be expressed in a way that students can understand? A teacher who can formulate and weigh these questions for herself is in a position to decide whether to use Josh's error to further everyone's learning, because she knows how to do it.

Those who prepare prospective teachers need to recognize how intellectually rich elementary-level mathematics is. At the same time, they cannot assume that these aspiring teachers have ever been exposed to evidence that this is so. Indeed, among the obstacles to improved learning at the elementary level, not the least is that many teachers were convinced by their own schooling that mathematics is a succession of disparate facts, definitions, and computational procedures to be memorized piecemeal. As a consequence, they are ill-equipped to offer a different, more thoughtful kind of mathematics instruction to their students.

Yet, it is possible to break this cycle. College students with weak mathematics backgrounds can rekindle their own powers of mathematical thought. In fact, the first priority of preservice mathematics programs must be to help prospective elementary teachers do so: with classroom experiences in which *their* ideas for solving problems are elicited and taken seriously, their sound reasoning affirmed, and their missteps challenged in ways that help them make sense of their errors. Teachers able to cultivate good problem-solving skills among their students must, themselves, be problem solvers, aware that confusion and frustration are not signals to stop thinking, confident that with persistence they can work through to the satisfactions of new insight. They will have learned to notice patterns and think about whether and why these hold, posing their own questions and knowing what sorts of answers make sense. Developing these new mathematical habits means learning how to continue learning.

The key to turning even poorly prepared prospective elementary teachers into mathematical thinkers is to work from what they *do* know—the mathematical ideas they hold, the skills they possess, and the contexts in which these are understood—so they can move from where they are to where they need to go. For their instructors, this requires learning to understand how their students think. The disciplinary habits of abstraction and deductive demonstration, characteristic of the way professional mathematicians present their work, have little to do with the ways each of us initially enters the world of mathematics, that is, experientially, building our concepts from action. And this is where mathematics courses for elementary school teachers must begin, first helping teachers make meaning for the mathematical objects under study—meaning that often was not present in their own elementary educations—and only then moving on to higher orders of generality and rigor.

The medium through which this ambitious agenda can be realized is the very mathematics these elementary teachers are responsible for—first and foremost, and still the heart of elementary content, number and operations; then, geometry, early algebraic thinking, and data, all of which are receiving increased emphasis in the elementary school curriculum.

This is not to say that prospective teachers will be learning the mathematics as if they were nine-year-olds. The understanding required of them includes acquiring a rich network of concepts extending into the content of higher grades; a strong facility in making, following, and assessing mathematical argument; and a wide array of mathematical strategies.

Below is a summary of the major themes in the areas of number and operations, algebra, geometry, and data to be addressed in the three courses for elementary teachers recommended in Chapter 2. Chapter 7 in Part 2 of this document, built around a set of classroom scenes, illustrates some of the central topics of the elementary curriculum, considers the content teachers must know in order to successfully manage the mathematical issues these scenes raise (an elaboration of the points below), and offers examples of the insights and struggles of teachers learning this content.

Number and Operations

To be prepared to teach arithmetic for understanding, elementary teachers, themselves, need to understand:

- A large repertoire of interpretations of addition, subtraction, multiplication and division, and of ways they can be applied.

- Place value: how place value permits efficient representation of whole numbers and finite decimals; that the value of each place is ten times larger than the value of the next place to the right; implications of this for ordering numbers, estimation, and approximation; the relative magnitude of numbers.

- Multidigit calculations, including standard algorithms, "mental math," and non-standard methods commonly created by students: the reasoning behind the procedures, how the base-10 structure of number is used in these calculations.

- Concepts of integers and rationals: what integers and rationals (represented as fractions and decimals) are; a sense of their relative size; how operations on whole numbers extend to integers and rational numbers; and the behavior of units under the operations.

The study of number and operations provides opportunities for prospective teachers to create meaning for what many had only committed to memory but never really understood. It should begin by placing the mathematics in everyday contexts—e.g., comparing, joining, separating, sharing, and counting quantities that arise in one's daily activities—and working with a variety of representations—e.g., number lines, area diagrams, and arrangements of physical objects. Instead of solving word problems by looking for "key words" or applying other superficial strategies, prospective teachers should learn to consider the actions the problems might posit. Learning to recognize that a single situation can be modeled by different operations opens up discussion of how the operations are related.

Future teachers must understand the conceptual underpinnings of the conventional computation algorithms as well as alternative procedures such as those commonly generated by children, themselves. (For example, in the vignette, Laurel, Chris, Jack, and Dava present various methods for calculating a product. An example of an addition calculation: a child explains her method for adding $58 + 24$

is "Take 2 from the 24 and add it to the 58 to make 60; $60 + 22 = 82$.") This process might begin by having teachers perform multidigit calculations mentally, without the aid of pencil and paper, to help loosen the hold of the belief that there is just one correct way to solve any mathematics problem. *As they become aware and then pursue their own ideas, they will recognize, often for the first time, that they do, indeed, have mathematical ideas worth following.* Similar exercises can be used to help teachers see how decimal notation allows for approximation of numbers by "round numbers" (multiples of powers of 10), facilitating mental arithmetic and approximate solutions.

Although most teachers are able to identify the ones place, the tens place, etc. and write numbers in expanded notation, they often lack understanding of core ideas related to place value. For example, future teachers should understand: how place value permits efficient representation of large numbers; how the operations of addition, multiplication, and exponentiation are used in representing numbers as "polynomials in 10"; and how decimal notation allows one to quickly determine which of two numbers is larger. Furthermore, they should be familiar with the notion of "order of magnitude."

Having developed a variety of models of whole number operations, teachers are ready to consider how these ideas extend to integers and rational numbers. First they must develop an understanding of what these numbers are. For integers, this means recognizing that numbers now represent both magnitude and direction. And though most teachers know at least one interpretation of a fraction, they must learn many interpretations: as part of a whole, as an expression of division, as a point on the number line, as a rate, or as an operator. Teachers may have learned rules for comparing fractions, but now, equipped with a choice of representations, they can develop flexibility in determining relative size.

As with whole number operations, placing operations with fractions in everyday contexts helps give meaning to algorithms hitherto regarded as mechanical devices. (Many college students see fractions only as pairs of natural numbers plugged into arithmetic procedures; hence, to them, adding two fractions is simply a computation with four integers.) Teachers must recognize that some generalizations often made by children about whole number operations, e.g., a product is always larger than its factors (except when a factor is 0 or 1) and a quotient is always smaller than its dividend (unless the divisor is 1) no longer hold, and that the very meanings of multiplication and division must be extended beyond those derived from whole number operations.

The idea of "unit"—that the same object can be represented by fractions of different values, depending on the reference whole—is central to work with fractions. In addition and subtraction, all the quantities refer to the same unit, but do not in multiplication and division.

Another area to be explored is the extension of place-value notation from whole numbers to finite decimals. Teachers must come to see that any real number can be approximated arbitrarily closely by a finite decimal, and they must recognize that the rules for calculating with decimals are essentially the same as those for whole numbers. Explorations of decimals lend themselves to work with calculators particularly well.

As with all of the content described in this document, the topics enumerated are not to be taught as discrete bits of mathematics. Always, the power comes from connection—using the concepts and skills flexibly, recognizing them from a variety of perspectives as they are embedded in different contexts.

Algebra and Functions

Although the study of algebra and functions generally begins at the upper-middle- or high-school levels, some core concepts and practices are accessible much earlier. If teachers are to cultivate the development of these ideas in their elementary classrooms, they, themselves, must understand those concepts and practices, including:

- Representing and justifying general arithmetic claims, using a variety of representations, algebraic notation among them; understanding different forms of argument and learning to devise deductive arguments.

- The power of algebraic notation: developing skill in using algebraic notation to represent calculation, express identities, and solve problems.

- Field axioms: recognizing commutativity, associativity, distributivity, identities, and inverses as properties of operations on a given domain; seeing computation algorithms as applications of particular axioms; appreciating that a small set of rules governs all of arithmetic.

- Functions: being able to read and create graphs of functions, formulas (in closed and recursive forms), and tables; studying the characteristics of particular classes of functions on integers.

Algebraic notation is an efficient means for representing properties of operations and relationships among them. In the elementary grades, well before they encounter that notation, children who are encouraged to recognize and articulate generalizations will become familiar with the sorts of ideas they will later express algebraically. In order to support children's learning in this realm, teachers first must do this work for themselves. Thus, they must come to recognize the centrality of generalization as a mathematical activity. In the context of number theory explorations (e.g., odd and even numbers, square numbers, factors), they can look for patterns, offer conjectures, and develop arguments for the generalizations they identify. And the arguments they propose become occasions for investigating different forms of justification. If, in this work, teachers learn to use a variety of modes of representation, including conventional algebraic symbols, the algebra they once experienced as the manipulation of opaque symbols can be invested with meaning.

Particularly instructive in work on word problems are comparisons of solution procedures using a variety of representations, illustrating how algebraic strategies mirror the actions modeled by other methods. As teachers become more confident of their skill in using algebra, they come to appreciate the advantages of its economy as against the cumbersomeness of other modes of representation, such as blocks or diagrams.

Although initially teachers' work in number and operations must be grounded experientially, now they are equipped to return to the study of computation, this time to appreciate the algorithms on whole numbers, integers, or rationals as applications of commutativity, associativity, distributivity, identities, and (when it holds) inverses, the small set of rules governing all of arithmetic.

Especially important for teachers is recognition of how young children's work with patterns can be related to the concept of function—for example, that labeling the terms or units of a pattern by the natural numbers creates a function. As they pursue the study of functions, teachers learn to move fluently among descriptions of situations, tables of values, graphs, and formulas. And as they explore, they become familiar with certain elementary functions on integers: linear, quadratic, and exponential. They also learn to work with functions defined by physical phenomena, say, distance traveled by a runner over time, growth of a plant over time, or the times of sunrise and sunset over a year.

Geometry and Measurement

For many years, the geometry curriculum for the elementary grades consisted of recognizing and naming basic two-dimensional shapes, measuring length with standard and non-standard units, and learning the formulas for the area and perimeter of a rectangle (and possibly a few other shapes). Because many students arrive in high-school geometry courses unprepared for its content, topics in geometry have recently been accorded a more prominent role in the curriculum of the lower grades. To most elementary teachers, their own encounter with high-school geometry notwithstanding, much of this material is new. In order to teach it to young children, they must develop competence in the following areas:

- Visualization skills: becoming familiar with projections, cross-sections, and decompositions of common two- and three-dimensional shapes; representing three-dimensional objects in two dimensions and constructing three-dimensional objects from two-dimensional representations.

- Basic shapes, their properties, and relationships among them: developing an understanding of angles, transformations (reflections, rotations, and translations), congruence and similarity.

- Communicating geometric ideas: learning technical vocabulary and understanding the role of mathematical definition.

- The process of measurement: understanding the idea of a unit and the need to select a unit appropriate to the attribute being measured, knowing the standard (English and metric) systems of units, understanding that measurements are approximate and that different units affect precision, being able to compare units and convert measurements from one unit to another.

- Length, area, and volume: seeing rectangles as arrays of squares, rectangular solids as arrays of cubes; recognizing the behavior of measure (length, area, and volume) under uniform dilations; devising area formulas for basic shapes; understanding the independence of perimeter and area, of surface area and volume.

A first goal in a geometry course for prospective teachers is the development of visualization skills—building and manipulating mental representations of two- and three-dimensional objects and perceiving objects from different perspectives. In exercises designed to cultivate these skills, teachers handle physical objects: build structures with cubes, create two-dimensional representations of three-dimensional objects, cut out and paste shapes. Such activities require that teachers work with projections and cross-sections, recognize rotations, reflections, and translations, and identify congruent parts.

From this work, teachers become familiar with basic two- and three-dimensional shapes: learn their names, learn to draw them, know their definitions and see how the shapes satisfy those definitions, recognize these shapes as parts of more complex configurations, and know some facts about them. They also develop different images of how shapes are composed: seeing a cube, say, as a stack of congruent squares or as an object whose surface unfolds into a net of six squares; or a tetrahedron as a stack of triangles decreasing in size or as an object whose surface unfolds into a net of four triangles.

In studying geometric shapes, teachers should cultivate technical vocabulary, developing an appreciation of the power of precise mathematical terminology as they work to communicate their ideas. Here, especially, the role of mathematical definition needs to be highlighted.

Prospective teachers are familiar with the concept of angle, but often only superficially. Teachers should understand the idea of angle, both as the figure formed by two rays sharing a vertex and as angular motion. They should understand that angles can be added, that the measure of the sum of angles is the sum of the measures (modulo 2π or 360 degrees), and that the measures of the angles of a triangle sum to 180 degrees (a straight angle); and be able to prove that the measures of the angles of an n-gon sum to $180(n-2)$.

Most prospective teachers understand the use of rulers, but few have had occasion to consider the conceptual issues involved in measurement. To measure an attribute, one must select a unit appropriate to that attribute, compare the unit to the object, and report the total number of units. Teachers should understand that measurements in the real world are approximations and that the unit used affects the precision of a measurement. They should be able to convert from one unit to another and be able to use the idea of conversions to estimate measure. In particular, teachers should know standard systems of units and approximate conversion rates from English to metric units and vice versa.

With regard to length, area, and volume, teachers should know what is meant by one, two, and three dimensions. (A common misunderstanding: perimeter is two dimensional since, after all, "the perimeter of a rectangle has both length and width.") Many teachers who know the formula A = L × W may have no grasp of how the linear units of a rectangle's length and width are related to the units that measure its area or why multiplying linear dimensions yields the count of those units. An understanding of the volume of a rectangular solid involves seeing the relationship between layers of three-dimensional units and the area of its base. Formulas for the area and volume of some other kinds of objects can build from an understanding of rectangles and rectangular solids. The study of rectangles and rectangular solids can also lead to an understanding of how length, area, and volume change under uniform dilation.

Data Analysis, Statistics, and Probability

Statistics is the study of data, and despite daily exposure to data in the media, most elementary teachers have little or no experience in this vitally important field. Thus, in addition to work on particular technical questions, they need to develop a sense of what the field is about. Prospective teachers need experience in:

- Designing data investigations: understanding the kinds of question that can be addressed by data, creating data sets, moving back and forth between the question (the purpose of the study) and its design.

- Describing data: understanding shape, spread, and center; using different forms of representation; comparing two sets of data.

- Drawing conclusions: choosing among representations and summary statistics to communicate conclusions, understanding variability, understanding some of the difficulties that arise in sampling and inference.

- Probability: making judgments under conditions of uncertainty, measuring likelihood, becoming familiar with the idea of randomness.

Teachers need to develop skill in the design and conduct of data investigations: to pose questions that can be addressed by data; design data collection procedures; collect and analyze those data; consider whether their initial questions have, indeed, been addressed; or, if necessary, revise both questions and data collection procedures and analyze the new data; and, finally, draw conclusions and communicate findings. Any of these steps can itself become an object of study. This includes understanding the kinds of questions that can be addressed by data; creating data sets; learning how to explore data through describing the shape, center, and spread of data distributions; and then using such descriptions to support conclusions. Teachers must have practice in analyzing the processes and causes of variability. In particular, they should understand that correlation does not imply causality.

In the early grades, children begin to explore the idea of making judgments under conditions of uncertainty; they talk about what is impossible or certain, more or less likely. Teachers should be able to extend these ideas to determine measures of likelihood: given equally likely outcomes, the probability of a particular event is equal to the ratio of the number of outcomes defined by the event to the number of total possible outcomes. A part of this study includes discussion of randomness and the difference between predicting individual events and predicting patterns of events.

Conclusion

Too many students preparing for elementary teaching have been less than successful mathematics students, and even those with good grades often doubt their competence. Understandably, readers of this document may feel dismay at the prospect of working with such math-anxious, if not math-phobic, undergraduates.

However, those who work with them can testify that, once these prospective teachers experience their own capacities for mathematical thought, their anxiety is transformed into energy for learning.

In taking responsibility for the kind of instruction for elementary teachers envisaged here, mathematicians are invited, in effect, to re-enter the world of the naïve mathematical thinker. The recognition that the "unsophisticated" questions teachers pose do raise fundamental issues should inspire instructors to find contexts in which these can be addressed fruitfully. This means, at least initially, approaching the mathematics from a concrete and experientially based, rather than an abstract/deductive, direction.

Chapter 4

Recommendations for Middle Grades Teacher Preparation

The more sophisticated content of middle grades[1] mathematics necessitates that mathematics specialists teach in these grades and that these specialists have a well-developed understanding of the mathematics they teach. The mathematics expected of middle grades students builds on but is qualitatively different from the mathematics of the earlier grades. Students, for example, are expected to be able to operate with understanding on rational numbers; measure not only geometric shapes but in a variety of situations; develop understanding of similarity, congruence, and symmetry; mentally operate on geometric shapes; recognize and represent linear relationships with tables, graphs, and equations; gather, appropriately represent, and interpret simple data sets; and apply basic concepts of probability to measure uncertainty. Students must develop the ability to recognize situations in which it is appropriate to use ratios rather than differences to make mathematical comparisons. They should be helped to see the connections that exist among the mathematical topics they learn.

Teachers of middle grades students must be able to build on their students' earlier mathematics learning and develop a broad set of new understandings and skills to help their students meet these more sophisticated goals. New curricula now being adopted in many districts reflect these goals and assume a high level of understanding on the part of the teacher. Teaching middle grades mathematics requires preparation different from, not simply less than, preparation for teaching high school mathematics, and certainly reflecting more depth than that needed by teachers of earlier grades (described in the previous chapter). Too few teacher preparation programs offer preparation targeted for teachers of the middle grades. In many cases these teachers have been prepared to teach elementary school mathematics and lack the broader background needed to teach the more advanced mathematics of the middle grades.

Recommendation 2 of Chapter 2 advocates a program of at least 21 semester-hours of mathematics for prospective mathematics teachers of these grades. Two types of courses should be included. First, courses must be designed that will lead prospective teachers to develop a deep understanding of the mathematics they will be teaching. The design of this coursework, approximately 12 semester-hours, is the focus of this chapter. Some of this coursework could overlap with coursework for

---

[1]The meaning of the term "middle school" varies by school district and at times even within districts where there are middle schools of grades 5–8, of grades 6–8, of grades 6–9, and other configurations. To alleviate this ambiguity, this report uses "middle grades" to refer to grades 5–8 and "secondary" to refer to grades 9–12.

K–4 teachers, particularly that concerning fundamental ideas, such as place value, that extend from whole numbers to decimals.

Second, courses are needed that will strengthen these prospective teachers' own knowledge of mathematics and broaden their understanding of mathematical connections between one educational level and the next, connections between elementary and middle grades as well as between middle grades and high school. This second type of coursework should be carefully selected from the options offered by the department, and would require a precalculus or college algebra background. One semester of calculus could be part of this second group of courses if there is (or could be designed) a calculus course that focuses on concepts and applications, as opposed to the traditional course offered to mathematics majors and engineers.

Number theory and discrete mathematics can offer teachers an opportunity to explore in depth many of the topics they will teach. A history of mathematics course can provide middle grades teachers with an understanding of the background and historical development of many topics in the middle grades curriculum. A mathematical modeling course, depending on the level and substance of the course, can provide prospective teachers with understanding of the ways in which mathematics can be applied. If the prospective teachers are likely to teach algebra, coursework in linear algebra and modern algebra would be appropriate. If, in addition, the teachers might be expected to teach a full-year course in geometry, then they should have the same geometry coursework as prospective secondary teachers. These options would most likely require more than 21 semester-hours.

Instructional Themes for Courses Designed for Prospective Teachers

Making sense of mathematics should be a cross-cutting theme throughout K–12 mathematics instruction and in courses for prospective teachers. For many prospective teachers learning mathematics has meant only learning its procedures and, they may, in fact, have been rewarded with high grades in mathematics for their fluency in using procedures. Although procedural fluency is necessary, it is not an adequate foundation for teaching mathematics. An orientation towards making sense of mathematics must be considered fundamental both to learning and to teaching mathematics.

Problems involving proportionality permeate middle grades curricula, and therefore provide a focus for much of the content discussed here. Proportions occur, for example, in arithmetic tasks involving ratios and percents, in geometry tasks involving similarity, in algebra tasks involving linearity, and in tasks requiring the assignment of a probability to an event. Many prospective teachers have had only a one- or two-day encounter with proportions in their own schooling. That encounter may have involved only what are often called missing-value problems intended to be solved by applying a particular procedure: setting up a proportion and cross-multiplying to find an unknown value. If they are limited to this understanding teachers may fail to recognize when tasks are proportional in nature and so will miss opportunities to help their future students develop ability to reason appropriately with proportions. Proportional reasoning is a psychologically and mathematically sophisticated form of reasoning based on intuitive preschool experiences and developed in school through appropriate experiences (Sowder et al., 1998). For example, this problem requires proportional reasoning, but goes well beyond the standard matching of two ratios and cross-multiplying:

> In a certain town, the demand for rental units was analyzed, and it was determined that, to meet the community's needs, builders would be required to build apartments in the following way: Every time they build three single-bedroom apartments, they should build four 2-bedroom apartments and one 3-bedroom apartment. Suppose a builder is planning to build a large apartment complex counting between 35 and 45 apartments. How many 1-bedroom apartments, 2-bedroom apartments, and 3-bedroom apartments would the apartment building contain?   (Lamon, 1993, p. 44)

Another theme of middle school mathematics is that of variables and relations, and especially the multiple equivalent representations of relations among variables. Through their work on variables and relations students begin to develop an understanding of functions and of the power of symbols. This initial work sets the stage for further work in mathematics. Teachers need to foster algebraic thinking (e.g., recognizing patterns and finding governing rules, being able to think about quantities without necessarily attaching values, and being able to reverse operations [Driscoll, 1999]) in the early grades and then more particularly in the middle grades. To do so they need to recognize what types of tasks lend themselves to helping their students begin to reason about variables, relations, and functions, tasks involving such activities as making tables, searching for patterns, graphing, and generalizing.

Explaining one's reasoning in carrying out mathematical tasks is difficult for most adults who have never been required to do so in previous mathematics coursework. But this ability is fundamental to successful instruction and should be required in courses designed for this population. When reasoning, explaining, and sense-making are emphasized in teacher preparation courses, prospective teachers are more likely to assimilate these ways of thinking and communicating and apply them to their own learning, and they will be better prepared to model these aspects of mathematics when they begin teaching.

The remainder of this chapter discusses four content areas of mathematics in terms of developing prospective teachers' understanding. This understanding is dependent on reasoning, explaining, and sense-making. Topics in these four areas are found throughout the K–12 curriculum but vary in emphasis at each of level of schooling. This chapter focuses on the fundamental concepts and skills needed to teach mathematics well in the middle grades.

Number and Operations

Coursework for prospective middle grades teachers should lead them to:

- Understand and be able to explain the mathematics that underlies the procedures used for operating on whole numbers and rational numbers.

- Understand and be able to explain the distinctions among whole numbers, integers, rational numbers, and real numbers, how they are placed on the number line, and how field axioms hold or do not hold depending on the types of numbers being used.

- Convert easily among fractions, decimals, and percents.

- Demonstrate facility in using number and operation properties, including mental computation and computational estimation.

- Understand and be able to explain fundamental ideas of number theory as they apply to middle school mathematics.

- Make sense of large and small numbers and use scientific notation.

- Apply proportions appropriately and provide explanations.

Number continues to be a focus of mathematics in the middle grades, but there is a shift in these grades from work with whole numbers to work with rational numbers. By the time students complete the middle grades, they should be able to recognize which arithmetic operation or operations can be used appropriately to solve a particular problem and to carry out the procedures to obtain a correct answer. They also should be able to use numbers in sensible ways such as recognizing when an answer is unreasonable; comparing fractions by recognizing their relative distances from 0, 1, or $1/2$; shifting easily among fractions, decimals, and percents; and knowing when a mental calculation would be faster than using paper and pencil or even a calculator. In curriculum materials, using numbers in sensible and flexible ways is often referred to as number sense.

Coursework designed for middle grades teachers must help these teachers develop strong number sense in addition to number skills. The prospective teachers will also need to learn to go beyond procedures to look at the underlying mathematics and try to make sense of it. One way to accomplish this goal is to provide them with examples of middle grades problems, and if possible, examples of students' work on these problems, and discuss the strengths and weaknesses of different students' approaches to a problem. (Books on results of mathematics portions of the National Assessment of Educational Progress and items from the 8th grade Third International Mathematics and Science Study are sources of such problems.[2]) For example, suppose seventh graders have been asked to estimate 316×16.2. Here are five possible responses:

- 4500, because 300×15 is 4500.

- Round 16.2 to 20 and round 316 to 300. My estimate is 6000, minus a little.

- You can round 316 to 320 and think of 320 as 32×10. 16.2 rounds to 16. Then you can write the problem as $2^5 \times 2^4 \times 10$, which is $2^9 \times 10$, which is 5120.

- 300×16 is 4800, 10×16 is 160 and 5×16 is 80. So the sum is 4800 plus 240 which is 5120.

- If you multiply 316 and 16.2, you get 5119.2, which rounds to 5100.

---

[2]Silver and Kenney's *Results from the Sixth Mathematics Assessment* is an example of the former. TIMSS items are available at `http://timss.bc.edu/TIMSS1/TIMSSPublications.html`.

Prospective teachers (who should have already made their own estimates) could then be asked about the mathematical thinking portrayed by each of these solutions. What type of understanding of numbers and operations is represented in each solution? Does proficiency at computational estimation require a different understanding of numbers and operations than is needed to carry out standard computational procedures? Are some procedures and final estimates more, or less, appropriate in different situations, such as when estimating the cost of renting graduation gowns at $16.20 each for 316 students?

Many students in middle grades will not yet have fully developed necessary procedural skills for operations on whole numbers. Middle grades teachers need to understand and be able to explain both the rationale for the steps in whole-number arithmetic algorithms, which build on place value knowledge, and the ways these algorithms are extended to decimal number operations. In particular, the multiplication and division algorithms for whole numbers and decimal numbers need to be understood so that they can be taught in ways that help students remember them without resorting to thoughtless, rote techniques and that serve as a foundation for later learning. Teachers should be able to recognize valid procedures that students sometimes invent to carry out calculations, and to reward those that are based on good number sense. Many times the traditional algorithms can build on students' invented procedures and thus lead students to a better understanding of the standard procedures.

A strong foundation in work with rational numbers is absolutely essential for teaching in the middle grades. Prospective teachers often think that they have this knowledge if they know the algorithms for operations, for example, to invert and multiply when they divide fractions. They are sometimes surprised to learn that there is something to understand about the usual algorithm for dividing fractions and that there are other, equivalent algorithms. Understanding division of fractions requires a deep understanding of what fractions are, and of what division means.

Teachers should understand how decimals extend the place value work from the earlier grades. They should be able to convert easily among fractions, decimals, and percents. They should understand why only repeating decimals can be converted to fractions, and why non-repeating decimals are not rational, thus leading to a discussion of irrational numbers. Their knowledge of positive rational numbers can then be extended to a study of negative rational numbers. Although prospective teachers will have some familiarity with operational properties, the rational number system is usually their first encounter with a field. Teachers should be able to develop, for example, Venn diagrams to represent the hierarchy of the different types of numbers: whole, integer, rational, irrational, and real, and how they are related.

Mental computation and estimation can lead to better number sense. Most middle grades students and some prospective teachers, when asked to use mental computation, will attempt to mentally undertake the pencil-and-paper algorithm with which they are familiar rather than use number properties to their advantage (e.g., using the distributive property to find 7×28 or the associative property to think of 7×28 as $7 \times 7 \times 4$, or 49×4). To help prospective teachers develop "rational number sense," tasks can be designed that include mentally ordering a set of rational numbers (e.g., 0.23, $5/8$, 51%, and $1/4$) using knowledge of number size; estimating the outcomes of rational number operations (e.g., $7/8 + 9/10$ must be a little less than 2 because each fraction is a little less than 1), and recognizing wrong answers (e.g., $2/3 \div 1/2$ cannot be less than 1 because there is more than one $1/2$ in $2/3$). Developing flexibility in working with numbers will take time, even for prospective teachers, because most have never been asked to think about numbers in these ways.

Basic number theory has a valuable role to play in contemporary middle grades mathematics and should have a role in courses designed for middle grades teachers. They should experience conjecturing and justifying conjectures about even and odd numbers and about prime and composite numbers. They should have a good grasp of the Prime Factorization Theorem and how it extends to algebra learning. The difficulty of finding the greatest common factor of two numbers can lead students to an appreciation of the efficiency of the Euclidean Algorithm.

Prospective teachers need to attach meaning to very large numbers that they see daily. Developing *benchmarks* for large numbers (e.g., calculating one's share of the national debt) can lead to a better sense of what these numbers mean. Examples of very small numbers can be found in middle grades science. The difficulty of writing, expressing, and calculating with very large numbers and very small numbers will lead prospective teachers to appreciate the structure and sophistication of scientific notation.

Finally, experiences using ratios as a means of comparison can lead prospective teachers to think about situations that are proportional in nature. For example, when prospective teachers are asked to compare the steepness of two ramps, some do so by comparing the differences between the heights and depths of the ramps rather than by comparing the ratios of these two quantities. A problem such as this one can lead to finding slopes of lines in coordinate systems and understanding what the slope means. Percents are, of course, ratios, and need to be presented as such.

Algebra and Functions

Prospective middle grades teachers should:

- Understand and be able to work with algebra as a symbolic language, as a problem solving tool, as generalized arithmetic, as generalized quantitative reasoning, as a study of functions, relations, and variation, and as a way of modeling physical situations.

- Develop an understanding of variables and functions, especially of different equivalent relationships between variables.

- Understand linearity and how linear functions can illustrate proportional relationships.

- Recognize change patterns associated with linear, quadratic, and exponential functions.

- Demonstrate algebraic skills and be able to give a rationale for common algebraic procedures.

Viewing algebra as both a symbolic language useful in mathematics and science and as a powerful tool for making sense of the world is central to prospective teachers' understanding of what algebra is. Only when teachers work with a variety of problems designed to develop this expanded view of algebra will they come to understand what they are required to know in order to teach algebra. Algebra viewed in these different ways subsumes its traditional role of developing the ability to work efficiently and appropriately with symbols.

One way to develop meaning in algebra is to highlight the manner in which algebra is generalized arithmetic, a language that encodes properties of arithmetic operations. A somewhat different way to think of algebra is as an extension of quantitative reasoning in arithmetic situations. If arithmetic word problems are solved by focusing on the quantities in a problem and determining relationships among these quantities before assigning any numerical values to the quantities, it is a reasonable next step to assign variables rather than numbers. Assigning variables to the quantities and setting up equations representing the relationships is then a formalization of reasoning quantitatively about the problem. However, this formalization is not always an easy one. Prospective teachers need practice on solving problems situated in realistic contexts through this type of analysis, which can also help them develop a deeper appreciation of the important role variables play in algebra.

Functions, relations, and variation all play important roles in school algebra. In middle grades these roles are prominent as students come to understand algebraic functions, particularly linear functions and the manner in which they can illustrate proportional relationships. Using qualitative graphs, that is, graphs that have no numbers but that "tell a story," can lead to a deeper understanding of functional relationships. For example, suppose a vase widens rapidly from its base, then has straight sides, then gradually tapers in until the top is congruent with the base. If water is poured into the vase at a steady rate, what will the time/height-of-water graph look like?

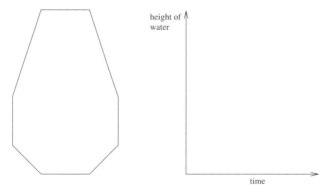

Examples such as this one can provide prospective teachers with a foundation needed to come to better understand how graphs can be used to represent linear, quadratic, and exponential functions, and how tables, equations, and graphs are related.

A study of the mathematics of change can provide a perspective on algebra that will be new to most prospective middle grades teachers. Middle grades teachers should know how to deal with problems involving rate of change, for example, of the growth of a plant, the decline of an endangered species, or of the speed of a car. The accumulation of miles traveled while driving at a certain speed is also a problem of change, and shows the relationship between distance traveled and rate of travel (Noble, Wright, Nemirovsky, & Tierney, 2001). Some elementary textbooks now include problems in which students investigate change. At the middle grades level, teachers who can provide examples of continuous change in situations involving time, velocity, and acceleration and who know how these changes can best be represented can lead students to a better understanding of how quantities such as time and motion are related. Some calculators now have as an accessory a motion detector that can be used to help prospective teachers better understand rate of change.

With technology and the advent of graphing calculators, there is much discussion about what students ought to be able to do with and without calculators. Teachers need to understand the potential that graphing calculators have for the enhancement of student learning of algebra; gaining this understanding requires that they use graphing calculators themselves.

Many curriculum developers say that the study of algebra ought to begin much earlier and extend throughout the grades, that algebra ought to be a curriculum strand rather than a course. For example, the study of patterns and making generalizations based on patterns can begin in the early grades and continue through the middle grades. Preparation of middle grades teachers ought to allow them to teach algebra as a curriculum strand or as a year-long course, depending on a particular school district's plan. To do either successfully, teachers must examine what school algebra is and how it can be taught in meaningful ways at different grade levels.

Measurement and Geometry

Prospective middle grades teachers should be able to:

- Identify common two- and three-dimensional shapes and list their basic characteristics and properties.

- Make conjectures about geometric shapes and then prove or disprove them.

- Demonstrate how rigid motions in the plane result in congruent figures.

- Demonstrate understanding of how similar figures result from a dilation, and the role of proportional relationships in determining similarity.

- Demonstrate ability to visualize and solve problems involving two- and three-dimensional objects.

- Connect geometry to other mathematical topics, and to nature and art.

- Understand the common forms of measurement and choose appropriate tools and units for measuring.

- Understand, derive, and use measurement techniques and formulas.

Prospective teachers have some basic knowledge about shapes and about how to calculate areas and volumes of common shapes, but many will not have explored the properties of these shapes or know why the area and volume formulas are true. The study of properties of shapes should include conjecturing, then proving or disproving to enhance geometric reasoning and lead to an understanding of the role of proof in geometry. The study of three-dimensional shapes beyond rectangular prisms is often new to prospective middle grades teachers, as is visualizing actions of shapes in three-dimensional space. Prospective teachers should develop spatial reasoning ability and should be able to perform such tasks as, for example, envisioning how to slice a cube to get a cross-section that is a square, or a non-square rectangle, or an equilateral triangle, or a trapezoid, or a regular hexagon. Software (e.g., *Geometer's Sketchpad*, *Cabri Geometry*, *Math Van*, *GeoShapes*, *Turtle Math*) can facilitate exploration of geometric shapes. Prospective middle grades teachers should have opportunities to become familiar with such software.

Making connections between geometry and other areas within mathematics is an important aspect of preparing teachers to teach mathematics. Reasoning with two- and three-dimensional shapes can lead quite naturally to a study of symmetry and of geometric transformations in general. The coordinate system so often used to represent equations in two variables lends itself well to investigating motions in the plane. The study of rigid motions can lead to an understanding of congruence, and dilations to an understanding of similarity, scale factors, and the role of proportions in establishing similarity. Geometry should also be studied as it occurs outside of mathematics, such as in nature and in art. There are many examples that could be studied, such as in the artwork of various cultures (see, e.g., Washburn and Crowe's *Symmetries of Culture* or Paulus Gerdes's *Geometry From Africa*). Geometric transformations can be found in many designs, and recognizing these transformations adds, for prospective teachers, a legitimacy to the study of transformations by middle grades students.

Prospective teachers need to understand what it means to measure—that a quantity is any characteristic (such as length or rate or number of eggs in a carton) that can be measured or counted using some preselected unit, that the measurement or count is the value of that quantity in terms of the given unit, and that standardized units are needed for ease of communication. Formulas for measuring area and volume should be developed in such a way that a teacher could later derive a formula if it is not remembered. But developing formulas is insufficient; teachers need to come to a deeper understanding of the quantity being measured. Many prospective teachers relate area only to geometric figures for which a formula can be derived and often use formulas inappropriately. They do not have strategies, such as decomposing and recomposing figures or superimposing grids, that lead to ways for finding areas of other shapes. Indeed, the idea that a non-regular closed figure in a plane has an area needs to be established for some prospective teachers. Part of their confusion stems from a lack of real understanding of area and of appropriate units to measure area. For example, if given a 3-inch by 5-inch card and asked to use it to measure the area of a table top, some prospective teachers use the card to measure one table edge in 5-inch lengths, measure the adjacent table edge in 3-inch lengths, and then are unable to identify a unit associated with the product of the two measurements.

Prospective teachers need experiences that help them better understand the role of units of measurement, choose appropriate tools and methods for measuring, and recognize the complexity of relationships between different types of measures. Exploring area and perimeter by holding one measurement constant is such an experience. Other learning situations could involve scale changes in planes and in space, leading first to problems involving proportional reasoning and measurements of similar figures, and then to the meaning of congruence of both two- and three-dimensional space. Other forms of common measurement, such as angle measurement, should be introduced in ways that help prospective teachers make sense of the manner in which these forms of measurement were devised and are used.

Measurement goes far beyond geometric figures. People have developed a variety of ways of measuring naturally occurring phenomena, from atmospheric pressure to the health of a new-born child, to satisfy the desire to quantify a particular quality. Dava Sobel's book *Longitude*, the basis of a 2000 PBS program, describes the need for accurately measuring time in order to find a ship's location at sea.

The Pythagorean Theorem and its applications in problems involving lengths of sides of triangles and in the development of the distance formula can fit well into a study of measurement. There are several proofs of the Pythagorean Theorem that are accessible to middle grades students. Teachers should have studied such proofs and be able to find, review, and use them in teaching when appropriate.

The discussion here focuses on general topics included in middle grades geometry and measurement. If they are to teach in schools where the usual secondary geometry course has been moved to middle grades, teachers will need further preparation, as discussed earlier.

Data Analysis, Statistics, and Probability

Coursework should provide prospective teachers avenues to:

- Design simple investigations and collect data (through random sampling or random assignment to treatments) to answer specific questions.

- Understand and use a variety of ways to display data.

- Explore and interpret data by observing patterns and departures from patterns in data displays, particularly patterns related to spread and variability.

- Anticipate patterns by studying, through theory and simulation, those produced by simple probability models.

- Draw conclusions with measures of uncertainty by applying basic concepts of probability.

- Know something about current uses of statistics and probability in many fields.

Prospective teachers need both technical and conceptual knowledge of the statistics and probability topics now appearing in middle grades curricula. Coursework for prospective teachers should include exploratory analyses of data sets, consideration of the various types of graphs that can be used to portray the data in ways that highlight the important features of the data, investigation of ways of measuring the center and spread of a distribution and how these measures affect decision-making, and the study of relationships portrayed by associations between two data sets (or two variables representing data). To make sense of the statistical information encountered in daily life (such as the accuracy of a political poll that declares a winner of an election or the correctness of a medical experiment that results in new treatment for a disease), a teacher must understand the role of the random selection of samples from a fixed and well-defined population for the purpose of estimating parameters of that population. Prospective teachers should understand the process of making inferences through simulated sampling distributions (which can be done effectively in middle grades) and also see how this process is related to the more mathematically based inference taught at higher levels.

Teachers need to understand that probability, because of its predictive value, can be considered a way of thinking about future events. For example, because the probability of heads is $1/2$ (or at least extremely close to $1/2$) on any toss of a fair coin, one should not bet heavily that the next toss will give tails, even though the last five tosses were heads. Knowledge that a model providing a theoretical probability can be checked empirically is fundamental to the study of the relative-frequency concept of probability, which is the most useful to the study of statistics at this level. Understanding probability distributions and how they arise leads students to better understand likely and unlikely outcomes. Summary measures for probability distributions are couched in terms of expected values, and expected-value problems should be included in a course for middle grades teachers. Many examples and much class discussion will be needed before prospective teachers can be expected to fully understand the rules for calculating probabilities of compound events made up of independent or dependent events.

Prospective teachers need experiences with designing simple experiments, collecting, displaying, and analyzing data, and using software that helps them understand how to display and interpret data. *Fathom* can be used to enhance teachers' learning; *DataScope* and *ProbSim* are available for Macs for use in the middle grades. *Tabletop* and a student version of *Data Desk* have been found useful by some teachers. New statistical software is being designed at University of Massachusetts to be used at the middle school level.[3]

All of this software can help both teachers and students develop conceptual understanding of statistical and probabilistic ideas. Access to appropriate computer technology is extremely important in helping prospective teachers to develop an understanding of statistical concepts.

Statistics and probability play increasingly important roles in many fields, and attention should be given to some of these roles so that prospective teachers understand the importance of this component of the curriculum.

Conclusion

The content described in this chapter is aimed at providing depth of understanding of the basic areas of middle grades mathematics. The preparation described here (and more extensively in Chapter 8) should result in teachers who

---

[3]For further information, see `http://www.umass.edu/srri/serg/index.html`.

know what it means to understand mathematics at a level that allows them to teach, and who have confidence in their own ability to engage their students in a strong core curriculum and prepare them academically for secondary school.

Chapter 5

Recommendations for High School Teacher Preparation

College and university programs for preparation of teachers often require mathematics coursework similar to that of a liberal arts mathematics major and education courses that emphasize teaching and learning of mathematics. Unfortunately, those programs do not appear to be attracting enough students to meet the national demand for new high school mathematics teachers. Furthermore, there is widespread concern that the mathematics courses in these programs do not provide prospective teachers with the depth and breadth of knowledge needed to teach high school mathematics well.

Recent recommendations for high school mathematics curricula and teaching demand even more of teacher preparation programs: that they provide their prospective teachers with knowledge of new mathematical topics, better understanding of the topics they will teach, and new teaching skills. In addition, future teachers need guidance in taking advantage of the increasingly sophisticated technological tools that permit more computationally involved applications and can give insights into theory. For example, computer software is now available that can be used to introduce high school students to discrete dynamical systems by iterating simple quadratic functions. The vignette that follows suggests the kind of mathematical breadth and depth that a high school teacher needs, illustrating new and longstanding concerns about teacher knowledge:[1]

> Ms. Liddell: The outside figure below is a 12 by 12 square. What is the area of the tilted square within it?

In response to this task two students came up with different answers. Reconciling their ideas led the class in quite unexpected directions.

> Julie: It looks to me as if the answer is 80. I wanted to see the rectangles—the ones the triangles are half of—so I drew the lines across and down like this:

---

[1]This vignette is based on material from Virginia Bastable's 1996 chapter "A Dialogue About Teaching."

Then I saw the tilted square was made up of four triangles and an even smaller square. The triangles are 16, which is half of 32. The little square is 4 by 4, so it is 16, too, for the four triangles and the little square, makes 80.

Bill: I got a different answer, but I see now that I didn't draw the figure right. I split the sides into sections of 3 and 9.

Alicia: Wouldn't the area be the same, anyway, no matter how you split up the sides?

Ms. Liddell: What do you think? If the point is moved from splitting the side of length 12 into sections of 4 and 8 to make other divisions, would the area of the tilted square stay the same?

While the students were considering that question, their teacher was thinking. She had to decide quickly whether to continue with her original plan—connecting the problem with irrational numbers—or to take advantage of the instructional opportunities raised by Bill's answer and Alicia's question: connecting the problem with the algebraic identity $(a + b)^2 = a^2 + 2ab + b^2$, asking her students for justification that the tilted figure is always a square, or to use a graphing calculator to explore the "tilted square area function" $y = 144 - 2(x)(12 - x)$.

In order to recognize the instructional opportunities described in the vignette, make a good decision, and implement it, a teacher needs a sound understanding of several areas of school mathematics and the connections among them. To recognize that Julie's diagram can be connected with $(a + b)^2 = a^2 + 2ab + b^2$ requires awareness of connections between algebra and geometry. To help students make this connection, a teacher needs the ability to identify particular correspondences between expressions and geometric objects. A teacher with an understanding of geometry and proof will notice that the implicit assertion in the problem statement that the inscribed figure is a square needs justification. A teacher with a knowledge of school curriculum will know if students have encountered geometric techniques which can be used for this justification. If students use a graphing calculator to explore the "tilted square area function" $y = 144 - 2(x)(12 - x)$, the teacher who recognizes that the graph of this function is a parabola, that it is symmetric, and that these facts have a geometric interpretation for this problem is better positioned to respond to the results of such explorations. All of these possibilities require a sound understanding of school mathematics, but an understanding that goes beyond the competence expected of a high school graduate. And, the need for this kind of understanding is not restricted to situations that require quick decisions. Teachers need a sound knowledge of mathematics to make curriculum decisions, plan lessons, and understand their students' work.

Current teacher preparation programs often do not explicitly focus on the kind of connections illustrated above for algebra, geometry, and functions—topics which are traditionally part of high school curricula—much less for newer topics such as data analysis or discrete mathematics. To meet these needs and to address the concerns discussed above, the education of prospective high school mathematics teachers should develop:

- Deep understanding of the fundamental mathematical ideas in grades 9– 12 curricula and strong technical skills for application of those ideas.

- Knowledge of the mathematical understandings and skills that students acquire in their elementary and middle school experiences, and how they affect learning in high school.

- Knowledge of the mathematics that students are likely to encounter when they leave high school for collegiate study, vocational training or employment.

- Mathematical maturity and attitudes that will enable and encourage continued growth of knowledge in the subject and its teaching.

This report recommends two main ways that mathematics departments can attain these goals. First, core mathematics major courses can be redesigned to help future teachers make insightful connections between the advanced mathematics they are learning and the high school mathematics they will be teaching. Second, mathematics departments can support the design, development, and offering of a capstone course sequence for teachers in which conceptual difficulties, fundamental ideas and techniques of high school mathematics are examined from an advanced standpoint. Such a capstone sequence would be most effectively taught through a collaboration of faculty with primary expertise in mathematics and faculty with primary expertise in mathematics education and experience in high school teaching.

This chapter outlines the skills and understandings that prospective high school teachers should acquire in their mathematical educations. These objectives are presented in five sections that correspond to major areas of high school curricula— algebra and number theory, geometry and trigonometry, functions and analysis, statistics and probability, and discrete mathematics—with important connections indicated where appropriate. Each section suggests ways in which mathematics major courses can provide useful learning experiences for prospective high school teachers and the kind of material that would be appropriate in capstone courses specifically for teachers. Each point in this chapter is elaborated in Chapter 9 in Part 2 of this report.

Algebra and Number Theory

The algebra of polynomial and rational expressions, equations, and inequalities has long been the core of high school mathematics. Current school mathematics curricula connect algebra to topics in functions and analysis, discrete mathematics, mathematical modeling, and geometry. Graphing calculators, spreadsheets, and computer algebra systems can encourage and facilitate those connections.

To be well-prepared to teach such high school curricula, mathematics teachers need:

- Understanding of the properties of the natural, integer, rational, real, and complex number systems.

- Understanding of the ways that basic ideas of number theory and algebraic structures underlie rules for operations on expressions, equations, and inequalities.

- Understanding and skill in using algebra to model and reason about real-world situations.

- Ability to use algebraic reasoning effectively for problem solving and proof in number theory, geometry, discrete mathematics, and statistics.

- Understanding of ways to use graphing calculators, computer algebra systems, and spreadsheets to explore algebraic ideas and algebraic representations of information, and in solving problems.

Calculus and linear algebra courses provide an opportunity to give undergraduates extensive practice with algebraic manipulation. Making this an explicit goal for these courses helps to assure that future teachers have technical "know how" in high school algebra. Upper division courses in abstract algebra and number theory examine mathematical structures that are the foundation for number systems and algebraic operations. These courses should assure that future teachers "know why" the number systems and algebra operate as they do. Unfortunately, too many prospective high school teachers fail to understand connections between those advanced courses and the topics of school algebra.

Prospective teachers can be helped to make these connections in courses for mathematics majors and the capstone sequence. Both number theory and abstract algebra courses can be infused with tasks that ask for specific instances of these connections, for example, to show explicitly how the number and algebra operations of secondary school can be explained by more general principles. Assignments in a number theory course might ask for the use of unique factorization and the Euclidean Algorithm to justify familiar procedures for finding common multiples and common divisors of integers and polynomials. Assignments in an abstract algebra course might ask for each step in the solution of a linear or quadratic equation to be justified by a field property; and to show how each extension of the number system, from natural numbers through complex numbers, is accompanied by new properties. To make connections of collegiate and high school mathematics more natural, it also makes sense to place greater emphasis in the college courses on the ring, integral domain, and field structures that are fundamental in high school algebra.

Algebraic connections between high school and college courses can be an explicit focus of the capstone sequence for teachers. For example, this sequence could profitably examine the historical evolution of key concepts in number theory and algebra and it could trace the development of key number and algebra ideas from early secondary school through contemporary applications. It could examine the crucial role of algebra in use of computer tools like spreadsheets and the ways that computer algebra systems might be useful in exploring algebraic ideas. Each facet in such a capstone treatment of number and algebra would provide teachers with insight into the structure of high school mathematics, its uses in science and technology or in the workplace, and the conceptual difficulties in learning number and algebraic concepts.

Geometry and Trigonometry

High school geometry was once a year-long course of synthetic Euclidean plane geometry that emphasized logic and formal proof. Recently, many high school texts and teachers have adopted a mixture of formal and informal approaches to geometric content, de-emphasizing axiomatic developments of the subject and increasing attention to visualization and problem solving. Many schools use computer software to help students do geometric experiments—investigations of geometric objects that give rise to conjectures that can be addressed by formal proof. Some curricula approach Euclidean geometry by focusing primarily on transformations, coordinates, or vectors; and new applications of geometry to robotics and computer graphics illustrate how mathematics is used in the workplace in ways that are accessible and interesting to high school students.

To be well-prepared to teach the geometry recommended for high school, mathematics teachers need:

- Mastery of core concepts and principles of Euclidean geometry in the plane and space.

- Understanding of the nature of axiomatic reasoning and the role that it has played in the development of mathematics, and facility with proof.

- Understanding and facility with a variety of methods and associated concepts and representations, including transformations, coordinates, and vectors.

- Understanding of trigonometry from a geometric perspective and skill in using trigonometry to solve problems.

- Knowledge of some significant geometry topics and applications such as tiling, fractals, computer graphics, robotics, and visualization.

- Ability to use dynamic drawing tools to conduct geometric investigations emphasizing visualization, pattern recognition, conjecturing, and proof.

Current calculus and linear algebra courses for mathematics majors often give college students extensive experience with important geometric ideas and representations—especially Cartesian and polar coordinates, vectors, transformations, and trigonometry. Typical teacher preparation curricula require an advanced geometry course that examines Euclidean and rudimentary non-Euclidean geometry from an axiomatic point of view. This collegiate geometry course is an opportunity for teachers to deepen their understanding of Euclidean facts and principles and their skill in use of careful axiom-based reasoning. But future teachers should also be exposed to 20th-century developments in geometry.

This can be accomplished in two ways. First, the goals and objectives of standard college geometry courses can be reconsidered and greater emphasis given to modern approaches. For instance, the geometry of congruence and similarity can be developed from axioms about isometries and similitudes, and these developments can be connected to the algebra of matrices and complex numbers. Second, geometry course offerings can be extended to include examples of new topics and tools. For instance, many high school and college students are intrigued by artistic and scientific problems in computer graphics (which can be connected with matrix algebra) and by geometric investigations using dynamic drawing tools such as *Cabri Geometry* or *Geometer's Sketchpad* (leading to conjectures that require proof or disproof). Both of these topics emphasize the important role of visualization in mathematics and both provide opportunities for teachers to strengthen their skills in using coordinates and representations.

Knowledge of geometry for teaching can also be provided in the capstone sequence. For example, this sequence can explicitly trace the historical development of key ideas, identifying and exploring questions that will be as difficult for students as for the mathematicians who first encountered them. It can examine the interplay of exploration and proof. Such a course might also be an ideal venue for re-examination of trigonometric and closely related geometric ideas—the laws of sines and cosines, identities, the Pythagorean Theorem, and similarity—to assure that prospective teachers have the depth of understanding that is essential for effective instruction.

Functions and Analysis

The concept of function is one of the central ideas of pure and applied mathematics. For nearly a century, recommendations for school curricula have urged reorganization of school mathematics so that study of functions is a central theme. Computers and graphing calculators now make it easy to produce tables and graphs for functions, to construct formulas for functions that model patterns in experimental data, and to perform algebraic operations on functions.

Prospective high school mathematics teachers need to acquire deep understanding of the concept of a function and of the most important classes of functions (polynomial, exponential and logarithmic, rational, and periodic). For functions of one and two variables teachers should be able to:

- Recognize patterns in data that are modeled well by each important class of functions.

- Identify functions associated with relationships such as $f(xy) = f(x) + f(y)$ or $f'(x) = kf(x)$ or $f(x + k) = f(x)$.

- Recognize equations and formulas associated with each important class of functions and the way that parameters in these representations determine particular cases.

- Translate information from one representation (tables, graphs, or formulas) to another.

- Use functions to solve problems in calculus, linear algebra, geometry, statistics, and discrete mathematics.

- Use calculator and computer technology effectively to study individual functions and classes of related functions.

Undergraduate mathematics majors encounter functions in calculus, linear algebra, and various elective courses. However, most acquire only procedural facility in using formulas involving functions for calculations, not a deep understanding of functions and related concepts like limits or continuity. Thus, it is important for prospective teachers to revisit the elementary functions of high school mathematics from an advanced standpoint, in much the same way that they revisit algebraic and number system operations.

This sort of reflective look at functions and their unifying role in mathematics could be a prominent part of the capstone course sequence. The capstone study of functions might examine again the role of computers as computational and graphing tools in mathematical work. Such activities might lead to examination of relationships between explorations and proof—as well as experience in the kind of complex problem solving required by mathematical modeling.

In traditional college preparatory curricula, the primary goal is preparing students for study of calculus. Calculus is now commonly taught in advanced placement form at many U. S. high schools, so it is now even more important that prospective high school teachers gain an understanding of calculus that will allow them to make informed decisions about content and emphasis in preparatory courses. However, a beginning teacher is not expected to be well-prepared to teach an advanced placement course like calculus. Those high school mathematics teachers who will assume responsibility for teaching advanced placement calculus will need additional content beyond their initial preparation coursework.

Data Analysis, Statistics, and Probability

Over the past decades, statistics has emerged as a core strand of school and university curricula. The American Statistical Association's Quantitative Literacy project has encouraged inclusion of data-driven mathematics curriculum modules. The College Board's advanced placement examination in statistics is attracting a substantial number of high school students. The traditional school mathematics emphasis on probability has evolved to include more statistics, often in the context of using data analysis to gain insight into real-world situations.

Curricula for the mathematical preparation of high school teachers should include courses and experiences that help them appreciate and understand the major themes of statistics. Teachers need experience in:

- Exploring data: using a variety of standard techniques for organizing and displaying data in order to detect patterns and departures from patterns.

- Planning a study: using surveys to estimate population characteristics and designing experiments to test conjectured relationships among variables.

- Anticipating patterns: using theory and simulations to study probability distributions and apply them as models of real phenomena.

- Statistical inference: using probability models to draw conclusions from data and measure the uncertainty of those conclusions.

- Technology: using calculators and computers effectively in statistical practice.

Moreover, probability has important applications outside of statistics. Thus, prospective teachers should also:

- Understand basic concepts of probability such as conditional probability and independence, and develop skill in calculating probabilities associated with those concepts.

Despite the production of very interesting statistics materials for schools, it has been hard to find room for the subject in curricula dominated by preparation for calculus. Although the new advanced placement test in statistics may help to change this situation, at present prospective high school mathematics teachers commonly come to undergraduate study with very little prior work in statistics. Most programs for mathematics majors allow little time for statistics until, at best, an upper division elective. At that point, mathematics majors often find themselves thrust into a calculus-based mathematical statistics course, and are likely to miss many fundamental ideas and techniques that are at the heart of high school statistics and probability.

Because statistics is first and foremost about using data to inform thinking about real-world situations, it is critical that prospective teachers have realistic problem-solving experiences with statistics. Because modern statistical work depends on extensive calculations, the prospective teachers should also be able to use statistical computation software to organize, display, and analyze complex data sets. It is essential to carefully consider the important goals of statistical education in designing courses that reflect new conceptions of the subject. Such courses will be appropriate for most mathematics majors, as well as prospective teachers. There are recommendations from the MAA's Committee on the Undergraduate Program in Mathematics that provide helpful guidance in design of these courses.

Discrete Mathematics and Computer Science

The increasing application of mathematical methods in disciplines outside of the physical and engineering sciences has stimulated development and use of several key topics in discrete mathematics. Topics in discrete mathematics now appear in high school curricula. High school students have found them as accessible as some more traditional topics and many find their applications engaging.

High school mathematics teachers should get exposure to ideas, methods, and applications in the following areas:

- Graphs, trees, and networks.

- Enumerative combinatorics.

- Finite difference equations, iteration, and recursion.

- Models for social decision-making.

In each of these areas prospective teachers should have experience working on applied problems arising from real-world situations. Teachers should learn to reason effectively with mathematical induction, which is especially important in discrete mathematics. As in the case of statistics, most prospective teachers come to undergraduate study with very limited prior exposure to discrete mathematics.

The emerging importance of discrete mathematics is driven by the pervasive uses of digital computers and by problems involved in design of computer hardware, algorithms, and software. Computers are now used as problem-solving and learning tools throughout high school mathematics and other disciplines, but the underlying principles of computer science are seldom treated explicitly in any current high school mathematics course.

Prospective high school mathematics teachers commonly enter undergraduate studies with facility in using computers for a variety of information processing tasks, but they do not know much of the underlying theory. Teacher preparation programs commonly include a computer science requirement, though the nature of that requirement has been evolving as rapidly as the field of computer science itself. Future teacher preparation programs should include work in computer science and related mathematics such as:

- Discrete structures (sets, logic, relations, and functions) and their applications in design of data structures and programming.

- Design and analysis of algorithms, including use of recursion and combinatorics.

- Use of programming to solve problems.

Conclusion

The preceding recommendations outline a challenging agenda for the education of future high school mathematics teachers. Mathematics departments may not be able to meet this challenge by creating new courses required solely of prospective high school mathematics teachers. However, many of the recommendations above are probably as appropriate for all mathematics majors as they are for prospective teachers. All students in abstract algebra courses can profitably re-examine their knowledge of elementary algebra. Mathematics majors can gain insight into axiomatic methods and proof in a college geometry course, and most will be interested in newer topics recommended for inclusion in the geometry course for teachers. The basic themes of statistics, discrete mathematics, and computer science that have been suggested for inclusion are consistent with CUPM recommendations over the past two decades for all mathematics majors.

In addition to suggesting some new topics and emphases in existing mathematics courses, this report recommends creation of a new capstone course sequence aimed especially at future teachers. This sequence is an opportunity for prospective teachers to look deeply at fundamental ideas, to connect topics that often seem unrelated, and to further develop the habits of mind that define mathematical approaches to problems. By including the historical development of major concepts and examination of conceptual difficulties, this capstone sequence connects individual mathematics courses with school mathematics and contributes to the mathematical understanding and pedagogical skills of teachers.

In light of the current severe shortage of qualified high school mathematics teachers it might seem foolhardy to recommend preparation that is more ambitious than current standards. One might argue that teacher preparation should focus on very thorough grounding in a few core subjects. However, current high school curricula cannot be taught successfully by teachers with such limited preparation. As mathematics departments work to develop the kinds of courses needed to provide better preparation for future high school teachers, those efforts will also be useful in work with in-service teachers as well. In both cases, a teacher's preparation needs to be viewed as a foundation for a career of continuing professional development.

Chapter 6

Recommendations for Technology in Teacher Preparation

Technology has revolutionized many jobs and substantially increased the mathematical skills needed across the workforce. In contrast, its impact on instructional practices has been more modest and varies greatly from classroom to classroom. During the careers of the prospective teachers now in college classes, new technologies now unimaginable are likely to enter classrooms. Hence prospective teachers' most important need with regard to technology in college mathematics classes is in building a framework that will aid them in teaching with technology throughout their careers.

This chapter briefly discusses three general aspects of technology's role in school mathematics classrooms: (i) computation, (ii) new computer tools, and (iii) mathematical content. More specific uses of technology in school mathematics instruction and in college mathematics courses for prospective teachers are described in Part 2 of this report. It is assumed that all prospective teachers will use the Internet to access information sources and use e-mail to foster professional development discussions. It is beyond the scope of this report to discuss various types of computer-based instruction and distance learning.

Computation. First, prospective teachers need to understand the differences among the three most common types of computation in mathematics classrooms: (a) electronic computation to advance learning, (b) human computation to advance learning, and (c) electronic computation as a practical expedient.

Electronic computation. Electronic computation allows students to explore more complicated applications than would be possible in its absence and to work problems with realistic data. Students can use technology to visually study mathematical patterns, e.g., comparing the graphs of $y = 2x$ and $y = x^2$, and to approximate the value of complex expressions, e.g., arc-length integrals. The pervasive use of statistics in the modern world would be impossible without computers, and likewise it is natural in the classroom to use technology for virtually all statistical computations.

Human computation. When elementary grade students are learning arithmetic, they need extensive pencil-and-paper experience to develop fluency and accuracy with basic arithmetic procedures. They also need experience doing mental calculations. These arithmetic skills provide a critical foundation for subsequent learning about decimals, fractions, algebra, and more.

Practical expedient. By high school and continuing throughout their lives, people use calculators (or mental arithmetic) as an expedient way to obtain answers in arithmetic calculations.

Second, prospective teachers need to understand that the use of technology for complicated computation does not eliminate the need for mathematical thinking but rather often raises a different set of mathematical problems. Students should be shown that brute force computation often takes excessive time, even with modern computers. Considerable mathematical planning and theoretical insights may be needed to structure efficient computation.

New computer tools. Spreadsheets and modern programming languages open up new avenues of problem solving and reasoning. There is a growing array of educational software, such as *Geometer's Sketchpad*, *FunctionProbe*, or *Fathom*, designed to enhance students' understanding of mathematical concepts and their reasoning skills. Mathematicians, mathematics education researchers, and mathematics teachers have been exploring how such computer tools can support new types of pedagogy.

Prospective teachers at all levels need experience in using spreadsheets, ideally in several quantitatively based courses across the undergraduate curriculum. Prospective high school teachers need experience writing in computer programs in a high-level language, such as C++. Prospective high school teachers should also have experience with a computer algebra system, dynamic geometry software, and a statistical software package. These experiences should be designed with two goals in mind: the short-term goal of helping prospective teachers to effectively use current technology in teaching; and the long-term goal of enabling them to become thoughtful and effective in choosing and using educational technology.

Mathematical content. The mathematical topics and reasoning underlying computer science and information technology seem likely to play a larger role in future school mathematics curricula. Computer-related mathematics involves new facets of traditional topics, such as number theory for encryption, as well as new fields, such as discrete mathematics. Moreover, technology-related workplace and scientific needs may cause curricular changes. For example, because of the important role that analytic geometry plays in computer graphics and visualization, more attention may be given in the future to two-dimensional analytic geometry, and three-dimensional analytic geometry may re-enter high school curricula. Computer-related mathematics also involves different emphases in reasoning, such as more definitions by recursion. Mathematics faculty need to stay abreast of changes as school mathematics curricula evolve to address the mathematical needs of computer science and quantitative social sciences, along with the longstanding needs of the physical sciences.

References

Ball, D. L. (1991). Research on teaching mathematics: Making subject matter knowledge part of the equation. In J. Brophy (Ed.), *Advances in research on teaching* (Vol. 2, pp. 1–48). Greenwich, CT: JAI Press.

Bastable, V. (1996). A dialogue about teaching. In D. Schifter (Ed.), *What's happening in math class?* (Vol. 1, pp. 45–65). New York: Teachers College Press.

Driscoll, M. (1999). *Fostering algebraic thinking.* Portsmouth, NH: Heinemann.

Ewing, J. (Ed.). (1999). *Towards excellence: Leading a mathematics department in the 21st century.* Providence, RI: American Mathematical Society.

Gerdes, P. (1999). *Geometry from Africa: Mathematical and educational explorations.* Washington, DC: Mathematical Association of America.

Kenney, P. A., & Silver, E. A. (1997). *Results from the Sixth Mathematics Assessment.* Reston, VA: National Council of Teachers of Mathematics.

Leitzel, J. R. C. (Ed.). (1991). *A call for change: Recommendations for the mathematical preparation of teachers of mathematics.* Washington, DC: Mathematical Association of America.

Lamon, S. J. (1993). Ratio and proportion: Connecting content and children's thinking. *Journal for Research in Mathematics Education, 24,* 41–61.

Ma, L. (1999). *Knowing and teaching elementary mathematics: Teachers' understanding of mathematics in China and the United States.* Mahwah, NJ: Lawrence Erlbaum Associates.

McKnight, C., Crosswhite, F., Dossey, J., Kifer, E., Swafford, J., Travers, K., & Cooney, T. (1987). *The under-achieving curriculum: Assessing U. S. schools from an international perspective.* Champaign, IL: Stipes.

National Council of Teachers of Mathematics. (1989). *Curriculum and evaluation standards.* Reston, VA: Author.

National Council of Teachers of Mathematics. (2000). *Principles and standards of school mathematics.* Reston, VA: Author.

National Science Foundation. (1999). *Investing in tomorrow's teachers: The integral role of two-year colleges in the science and mathematics preparation of prospective teachers.* Washington, DC: Author.

Noble, T., Nemirovsky, R., Wright, T., & Tierney, C. (2001). The mathematics of change in multiple environments. *Journal for Research in Mathematics Education, 32*(1), 85–108.

Schifter, D., Bastable, V., Russell, S. J. (with Cohen, S., Lester, J. B., & Yaffee, L.) (1999). *Building a system of tens: Casebook.* Parsippany, NJ: Dale Seymour Publications.

Sobel, D. (1995). *Longitude.* New York, NY: Walker and Company.

Sowder, J., Armstrong, B., Lamon, S., Simon, M., Sowder, L, & Thompson, A. (1998). Educating teachers to teach multiplicative structures in the middle grades. *Journal of Mathematics Teacher Education, 2,* 127–155.

Stevenson, H., & Stigler, J. (1992). *The learning gap.* New York: Simon & Schuster.

Stigler, J., & Hiebert, J. (1999). *The teaching gap.* New York: Free Press.

Washburn, D. K., & Crowe, D. W. (1988). *Symmetries of culture: Theory and practice of plane pattern analysis.* Seattle: University of Washington Press.

Appendix

Relevant Reports

American Council on Education. (1999). *To touch the future: Transforming the way teachers are taught.* Washington, DC: Author. Available at `http://www.acenet.edu`.

Commission on Teaching and America's Future. (1997). *Doing what matters most: Investing in quality teaching.* New York: National Commission on Teaching & America's Future.

Committee on Science and Mathematics Teacher Preparation. (2000). *Educating teachers of science, mathematics, and technology: New practices for the new millennium.* Washington, DC: National Academy Press. Available at `http://books.nap.edu/catalog/9832.html`.

Ewing, J. (Ed.). (1999). *Towards excellence: Leading a mathematics department in the 21st century.* Providence, RI: American Mathematical Society. Available at `http://www.ams.org/towardsexcellence`.

International Society for Technology in Education. (2000). *National educational technology standards for teachers.* Eugene, OR: Author.

Kilpatrick, J., Swafford, J. & Findell, B. (Eds.). (2001). *Adding it up: Helping children learn mathematics.* Washington, DC: National Academy Press. Available at `http://www.nap.edu/catalog/9822.html`.

Learning First Alliance. (1998). *Every child mathematically proficient.* Washington, DC: Author. Available at `http://www.learningfirst.org/publications.html`.

Leitzel, J. R. C. (Ed.). (1991). *A call for change: Recommendations for the mathematical preparation of teachers of mathematics.* Washington, DC: Mathematical Association of America.

National Commission on Teaching and America's Future. (2000). *Before it's too late: A report to the nation from the National Commission on Teaching and America's Future.* Washington, DC: U.S. Department of Education. Available at `http://www.ed.gov/americacounts/glenn`.

National Council of Teachers of Mathematics. (2000). *Principles and standards of school mathematics.* Reston, VA: Author. Available at `http://www.nctm.org`.

National Science Foundation. (1999). *Investing in tomorrow's teachers: The integral role of two-year colleges in the science and mathematics preparation of prospective teachers.* Washington, DC: Author. Available at `http//www.nsf.gov`.

PART 2

Chapter 7

The Preparation of Elementary Teachers

The power to reason mathematically is a natural human capacity. Young children enter school already curious about number and size, and with ideas about how to join, remove, and split quantities. Mathematics instruction in the elementary years can—*should*—be designed to cultivate this curiosity. Encouraged to solve problems, children become aware of their ideas; and as they learn to analyze their own, their classmates, and their teachers thinking, these ideas become more refined and many-sided. It is during these early years that young students lay down those habits of reasoning upon which later achievement in mathematics will crucially depend.

Teaching elementary mathematics requires both considerable mathematical knowledge and a wide range of pedagogical skills. For example, teachers must have the patience to listen for, as well as the ability to hear, the sense—the logic—in children's mathematical ideas. They need to see the topics they teach as embedded in rich networks of interrelated concepts, know where, within those networks, to situate the tasks they set their students and the ideas these tasks elicit. In preparing a lesson, they must be able to appraise and select appropriate activities, and choose representations that will bring into focus the mathematics on the agenda. Then, in the flow of the lesson, they must instantly decide which among the alternative courses of action open to them will best sustain productive discussion.

It is by now widely acknowledged that many practicing teachers were not adequately prepared by the mathematics instruction they received to meet these challenges. As K–12 students—often even in the primary grades—they lost their curiosity about mathematics. When the rules and procedures one is taught are not conceptually anchored, memorization must pass for understanding, and mathematics becomes an endless, senseless parade of disparate facts, definitions, and procedures.

College students who today choose to become teachers are by and large still products of such K–12 instruction. Even if those who opt to teach middle- or high-school-level mathematics have experienced their mathematics education positively, many who choose elementary teaching have not. Intimidated by mathematics, the latter generally avoid mathematics courses wherever possible.

It seems, then, that we are caught in a vicious cycle: poor K–12 mathematics instruction produces ill-prepared college students, and undergraduate education often does little to correct the problem. Indeed, some universities mandate next to no mathematics coursework for the prospective elementary teacher. However, simply increasing the number of required credit hours is no solution—courses that allow students to get by using the same stratagems that got them through K–12 just perpetuate the problem.

In order to break this cycle, college students with weak mathematics backgrounds must have opportunities to reconnect with their own capacities for mathematical thought. Those among them who decide to enter the classroom and are willing to engage the conceptual riches of the elementary curriculum can become effective mathematics teachers. But they, just like the children they will someday teach, must have classroom experiences in which they become reasoners, conjecturers, and problem solvers.

Future teachers will need to connect fundamental concepts to a variety of situations, models, and representations. They will have to learn to notice patterns and think about why those patterns hold; pose their own questions and know what sorts of answers make sense; look for connections among different methods for solving the same problem or different ways of representing the same quantity. In short, developing these new mathematical habits means learning how to continue learning.

This is a daunting agenda. But if teachers are to help their students become strong mathematical thinkers, it must be met. And the medium through which this agenda can be realized is the very mathematics they are charged with teaching in the realms of number and operations, geometry, early algebraic thinking, and data.

Conventional belief has it that elementary school mathematics is simple and to teach it requires only learning prescribed facts and computational algorithms. However, recent work has revealed the conceptual richness of this early content, demonstrating that teaching elementary school mathematics can be intellectually challenging. Though each of us once inhabited the mathematical world of the young child, that world is lost to most of us. To re-enter it, Deborah Ball and Hyman Bass argue in a recent paper,

> one needs to be able to deconstruct one's own mathematical knowledge into less polished and final form, where elemental components are accessible and visible. We refer to this as *decompression*. Paradoxically, most personal knowledge of subject matter knowledge, which is desirably and usefully compressed, can be ironically inadequate for teaching. In fact, mathematics is a discipline in which compression is central. Indeed, its polished, compressed form can obscure one's ability to discern how learners are thinking at the roots of that knowledge. Because teachers must be able to work with content for students in its growing, not finished state, they must be able to do something perverse: work backward from mature and compressed understanding of the content to unpack its constituent elements.   (2000a, p. 98)

Ball and Bass's description of the challenge the elementary teacher faces in connecting to the mathematical world of the child holds for the instructor of the mathematically naïve adult. In a college course, prospective teachers' ingenuous questions will require instructors to "decompress" *their* mathematical knowledge to find responses satisfying to both mathematician and teacher.

Although some questions elementary-school teachers pose may be stimulating, others are certain to be very disturbing. Instructors teaching teachers for the first time will occasionally feel dismay, or even shock. How can such basic notions *not* be understood? What is there to think about? But the gaps in these teachers' mathematical backgrounds are consequences of systemic rather than personal failings, and it is essential that, recalling this, instructors work to maintain the necessary stance of interest, generosity, and respect.

To repeat, the challenge is to work from what teachers *do* know—the mathematical ideas they hold, the skills they possess, and the contexts in which these are understood—so they can move from where they are to where they need to go. For their instructors, as we have seen, this means learning to understand how their students think. The habits of abstraction—of compression—and deductive demonstration, characteristic of the way mathematicians present their work, have little to do with the ways children build their mathematical world, experientially, modeling concepts on actions—counting out, dividing up, comparing heights or ages. . . . Mathematics courses for teachers must aim, first of all, at helping them develop ways of giving meaning to the mathematical objects under study, only then moving on to higher orders of generality and rigor.

Chapter 3 outlines the mathematics content teachers need to know for the K–4 classroom. What follows in this chapter expands upon that discussion. Precisely because what goes on in the elementary classroom will seem alien to many readers of this document, vignettes drawn from actual lessons are used to elucidate the issues. In these scenes, children articulate their mathematical thinking, showing their teachers what they understand and where they are confused. The scenes do not exhaust the territory, but they are representative of the mathematical issues that arise in a typical classroom when mathematics teaching is organized to elicit and build upon children's thinking. (Some scenes are taken from grade 5 classrooms, but are included in the belief that what comes up in a fifth grade class is likely to come up for the fourth grade teacher, too.) The vignettes are followed by discussion of the mathematics the teachers will need in order to identify the sense in their students thinking, know when a key mathematical idea is being missed, or anticipate when significant mathematical territory is being broached. What mathematical knowledge will help teachers navigate these situations in ways that support building stronger mathematical conceptions?

These scenes taken from classrooms are *not* intended as models of exemplary teaching. They have been chosen, not for emulation, but to illustrate the kinds of knowledge and skills required of elementary teachers. Because lessons in which children practice routine procedures typically do not present mathematical challenges to teachers, they are not included in this document. Their absence is not meant to imply that such activities have no place in the elementary mathematics classroom.

For readers who have little contact with aspiring or practicing elementary-level teachers, excerpts from teachers learning journals and episodes from teacher education courses are also included. These are intended to communicate what teachers themselves report as new mathematical insights.

All of the vignettes are drawn from actual classrooms. Many of the scenes of elementary-level lessons are paraphrases of cases written by the teachers themselves. Others are taken from videotape or records of classroom observations. Scenes from courses for teachers and excerpts from teachers journals are based on published literature, unpublished field notes, and personal communications.

The recommendations for course content draw on research about teacher and student knowledge. They also follow from the assumption that most teachers (though certainly there are exceptions) have had few, if any, opportunities to learn content that is just now entering the elementary curriculum, particularly topics in early algebra, geometry, and statistics.

This chapter discusses teaching and learning at both elementary-school and teacher-preparation levels. To minimize confusion, the term "children" is used to refer to elementary school students, and "teachers" refers to both practicing and aspiring teachers.

Number and Operations

Understanding number and operations and developing proficiency in computation have been and continue to be the core concerns of the elementary mathematics curriculum. Although almost all teachers remember traditional computation algorithms, their mathematical knowledge in this domain generally does not extend much further. Indeed, many equate the arithmetic operations with the algorithms and their associated notation. They have little inkling of how much more there is to know. In fact, in order to interpret and assess the reasoning of children learning to perform arithmetic operations, teachers must be able to call upon a richly integrated understanding of operations, place value, and computation in the domains of whole numbers, integers, and rationals.

Summary of number and operations content.

- Understanding models and interpretations of operations with whole numbers (i.e. the set of non-negative integers):

    - having a large repertoire of interpretations of addition, subtraction, multiplication and division, and of ways they can be applied.
    - understanding relationships among operations.

- Developing a strong sense of place value in the base-10 number system:

    - understanding how place value permits efficient representation of number.
    - recognizing the value of each place as ten times larger than the value of the next place to the right and the implications of this for ordering numbers and for estimation and approximation.
    - seeing how the operations of addition, multiplication, and exponentiation are used in representing numbers.
    - recognizing the relative magnitude of numbers.

- Understanding multidigit calculations, including standard algorithms, "mental math," and non-standard methods commonly created by students:

    - recognizing how the base-10 structure of number is used in multidigit computations.
    - recognizing how decimal notation allows for approximation by "round numbers" (multiples of powers of 10).

- recognizing the properties of commutativity, associativity, and distributivity as useful tools for organizing thinking about computation.
- developing flexibility in mental computation and estimation.

- Developing the concepts of integer and rational number and extending the operations to these larger domains:

  - understanding what integers are and the meaning of sign and magnitude.
  - understanding what rational numbers are, understanding fractions and decimals as representations of rationals, and developing a sense of their relative size.
  - knowing interpretations and applications for the arithmetic operations in the extended domains.
  - understanding the relationship between fractions and the operations of multiplication and division.
  - understanding how whole number arithmetic extends to integers and rational numbers.
  - understanding how any number represented by a finite or repeating decimal is rational, and conversely.
  - understanding how and why whole number decimal arithmetic extends to finite decimals and, in particular, how place value extends to decimal fractions.

In order to begin to explore the mathematics content knowledge required for teaching at the elementary level, consider the classroom scene below, which captures children at work on subtraction.

> **Scene 1, from a second grade classroom:** The children have been finding the difference between Jorge's height, 62″, and the height of Cinthia's little brother, Paulo, 37″. (Currently they are using inches so that the heights will be two-digit numbers. Later they will use centimeters to get three-digit numbers.) Many of the children use dots and ten-sticks to represent two-digit numbers.

> Gabriella: (She has drawn three dots, then two ten-sticks, then two dots, and written 25.) I said, "How much does Paulo have to grow?" so 37 plus 3 more (pointing to three dots) is 38, 39, 40, and 50 (pointing to a ten-stick), 60 (pointing to another ten-stick), 61, 62 (pointing to two dots). So this is 23 (gesturing to the three dots and two ten-sticks), 24, 25 more he has to grow to catch up with Jorge.

> Roberto: I shrunk the big guy down by taking away the little guy from him (gesturing to his drawing of the little guy beside the big guy and the line he drew across from the top of the little guy to the big guy). So 62 minus 37 is 25. I took three tens from the six tens and seven from the ten. That leaves three and these two are five and two tens left is 25.

Josué:	I did it like Gabriella but I wrote three and then my ten-sticks and two and then added them to get 25 more the little guy needs.
Ms. Lo Cicero:	Can someone else say in their words how Josué did it?
Nanci:	He used numbers and sticks to go 37 plus 3 is 40 plus 2 tens is 60 plus 2 to get to Jorge. So 2 tens and 5 is 25.
Ruffina:	I just counted in my mind 37, 47, 57, that's 20, then 58, 59, 60, 61, 62, so that's 5. 25.
María:	I subtracted Paulo from Jorge like Roberto did, but I used numbers. I took one of the tens to get enough to take away the seven so that was three and two more was five ones, and there were two tens left so 25.
Ms. Lo Cicero:	Can someone else tell how Roberto's and María's methods are alike?
Carlos:	They both took away the little guy to get the difference. They took away 37 from 62.
Ms. Lo Cicero:	Anything else?
Jazmin:	They both had to open a ten because there weren't seven ones to take away. So Roberto took his seven from that ten-stick. (Teacher points to show the ten-stick Roberto separated into seven and three, and looks questioningly at Jazmin.) Yes, there he took seven and left three. And María took a ten from the six tens and wrote it with the ones and then took the seven to leave three.
Ms. Lo Cicero:	So they were both thinking about taking ones from a ten but they wrote it in different ways?
Several students:	Yes.
Letticia:	And we know other ways to write subtraction, too.
Ms. Lo Cicero:	Yes, you have lots of ways you show taking away and comparing, too. Whose heights should we compare next?

(based on Hiebert et al., 1997, pp. 153–155)

Formed by an education equating mathematical strength with computational proficiency, too many teachers have been left with an impoverished understanding of the number system. To orchestrate a classroom discussion like the one above, or those presented below, teachers must be able to do more than demonstrate remembered procedures—for example, they must be able to select problems that anticipate the issues their students will next need to confront, and then assess whether what the children make of those problems advances the mathematical agenda. Such skills require much deeper understanding of number and operations than most teachers now hold. The bulleted items discussed below identify key points of entry into the mathematics of the elementary grades.

• *Each operation can model a variety of actions or situations.*

For years, elementary textbooks have suggested that teachers teach their students to solve word problems by finding "key words": "altogether" means add, "left" means subtract, etc. Beyond such superficial clues, many teachers associate each operation with just one possible action: joining with addition; taking away with subtraction; repeated addition with multiplication; and either finding the number of groups of a given size or, given the number of groups, finding the size of each group (usually not both) with division (Graeber & Tanenhaus, 1993). Hampered by their own limited understanding of the operations, teachers have had little more to offer their students. What is required is a sense of the different kinds of situations that can be modeled by each of the four basic operations.

In the classroom scene above, the teacher has selected a problem involving comparison of heights. By thinking through the situation, the children develop different solution strategies: Gabriella considers how much Paulo has to grow; Roberto thinks about "shrinking the big guy down."

A teacher who is aware of the range of situations that can be modeled by subtraction can consciously choose problems that promote a variety of ways of thinking about the operation. The following is an example of a common "take-away" problem: *Sue Ellen had 62 cents and then bought an item for 37 cents. How much money did she then have?* A second type of subtraction problem involves joining, with the starting quantity unknown: *Manny went to the beach yesterday and picked up 37 shells for his shell collection. He now has 62 shells. How many shells did he have before his last visit to the beach?*

Teachers must come to recognize the variety of situations—of joining, separating, and comparing, with an unknown in various positions—that can be represented by addition and subtraction. Similarly, multiplication and division can be associated with a rich store of interpretations: multiple groups, splitting, shrinking and stretching, counting rectangular arrays, counting combinations. In many interpretations of multiplication (and in contrast to addition and subtraction), the numbers are associated with different units: e.g., 3×4 might model 3 *bags*, each with 4 *donuts*. Such multiplication problems have two types of division problem as analogs: partitioning into groups of a given size and partitioning into a given number of groups.

In developing more broadly based conceptions of the kinds of situations modeled by the operations, it is also important to become familiar with such other modes of representation as the number line or arrangements of blocks. Area representations of multiplication are particularly useful.

Keeping in mind Scene 1, consider a second, in which, at the start of a new unit on division, the teacher has given the class a set of what she considers division problems.

Scene 2, from a combined third/fourth grade classroom:

Jesse has 24 shirts. If he puts eight of them in each drawer, how many drawers does he use?

Vanessa writes: $24 - 8 = 16$, $16 - 8 = 8$, $8 - 8 = 0$, and then writes 3 for the answer.

If Jeremy needs to buy 36 cans of seltzer water for his family and they come in packs of six, how many packs should he buy?

This time Vanessa writes: $6 + 6 = 12$, $12 + 12 = 24$, $24 + 6 = 30$, $30 + 6 = 36$. (She doesn't identify her answer.)

You go into a pet store that sells mice. There are 48 mouse legs. How many mice are there?

Matthew organizes his work in a chart of two columns:

$$1\,m \quad 4\,l$$
$$2\,m \quad 8\,l \ldots$$

Then in a neat box he writes,

$$12m \times 4l = 48l$$

Above the box, he writes the number 12.

The teacher wonders, what does this say about kids' understanding of division if they use all the operations *except* division?

(based on Schifter et al., 1999a, pp. 55-57)

Scenes 1 and 2 illustrate another issue, the interrelationships of the operations, that many teachers need to work on.

- *A given situation can be modeled by different operations.*

The "key words" mindset leads many teachers to believe that, for any given word problem, there is just one operation that can be used to solve it correctly. However, as Scene 2 shows, a "division problem" can be solved by adding, subtracting, or multiplying. And in Scene 1, where some children readily see their way to the solution as a process of finding a missing addend, others subtract. Instead of ruling out any of these methods as incorrect or problematic, a teacher who understands the ways operations are interrelated can seize the opportunity to explore such connections more deeply.

The children whose classroom gave rise to Scene 2 did eventually learn their division facts. With a richer understanding of what division means and how it is related to the other operations, they were able to see how particular facts can be derived from other facts, making the process of recall easier. (Later in the lesson from which Scene 2 derives, Matthew is given another "division problem." He responds, "It's $63 \div 9$. What number times 9 is 63? Seven. . . . [I]t is [division], but my thinking is multiplication.")

Understanding how the operations are related and how these relationships can be called upon in solving problems is critical for teachers if they are to interpret and advance their students thinking. Just such a revelation is recorded in this excerpt from the journal of one participant in an inservice course.

In my group we did it [$159 \div 13$] the "regular way" [using the division algorithm], then by equally distributing base-ten rods, [and then] by going around and counting out by ones. Then [the instructor] came to our group and $159 \div 13$ suddenly became $159 - 13 = 146$, $146 - 13 = 133$, $133 - 13 = 120$, . . . and so on. I had never before thought of division as directly related to subtraction. . . . As simple as it sounds that one interaction really made an impact on me, as $\times$ and $\div$ were just something

> I did by a rote method, with not much thought as to how $+$, $-$,
> $\times$, and $\div$ are all related.   (Schifter et al., 1999b, p. 177)

This teacher was now resolved to bring these ideas to her fourth grade students.

Returning to Scene 1, consider that the children's methods rely on decomposing numbers into tens and ones. This highlights another set of ideas teachers must understand.

- *The principles of place value involve significant conceptual issues for young children and for teachers.*

Most teachers are readily able to identify the ones place, the tens place, etc., and can represent multidigit numbers in expanded notation. Nonetheless, they often lack understanding of core ideas: how place value permits efficient representation of numbers; that the value of each place is ten times larger than the value of the next place to the right; how a number can be decomposed into tens and ones in a variety of ways (53 can be viewed as 5 tens and 3 ones, or 4 tens and 13 ones, etc.); how the operations of addition, multiplication, and exponentiation are used in representing numbers as "polynomials in 10"; and how decimal notation allows one to determine quickly which of two numbers is larger. Teachers should be familiar with the notion of "order of magnitude" and should have a sense of the relative magnitudes of numbers.

Not only must teachers be able to state these ideas, they must be able to recognize and apply them flexibly. One activity that has been used successfully to help teachers develop such facility asks them to create a number system using the symbols A, B, C, D, and 0 (Schifter & Fosnot, 1993). The letters allow the possibility of assigning different values to the symbols. Working in small groups and offered a set of base-five blocks as a thinking tool, teachers are asked to show how to represent large numbers in their system and to calculate with multidigit numbers. If they should get that far, they are also asked to explore divisibility and give an account of numbers smaller than 1.

Among the strategies teachers tend to pursue, the following are the most common:

- assigning the values 1, 5, 25, 125 to A, B, C, D; one counts A, AA, AAA, AAAA, B, BA, BAA, . . .

- assigning the values 1, 2, 3, 4, 5 to A, B, C, D, 0; one counts, A, B, C, D, 0, 0A, 0B, 0C, . . . (some teachers, uncomfortable assigning 5 to the symbol "0," make up a different fifth symbol).

- assigning the values 1, 2, 3, 4, 0 to A, B, C, D, 0; one counts, A, B, C, D, A0, AA, AB, . . .

The first two strategies result in number systems resembling those of the ancient Egyptians and the ancient Greeks, respectively; the third results in a place-value system.

Although it is intended that everyone eventually explore a base-five place-value system, getting there as quickly as possible is not the point of the exercise. More important is that the teachers suggest a system, explore it, encounter its limitations, and redesign it accordingly. Through this process, they discover the various properties of different number systems and gain deeper understanding of our own.

Many teachers exposed to this activity come to appreciate the kind of flexible, connected knowledge that allows them to recognize a familiar mathematical idea in a very unfamiliar setting. As one teacher wrote:

> I've taught place value over and over and over again and I've told the kids, "We only have ten numerals and the way the number system works is, the place tells you the value of the number." I've said it a hundred times and here I went to design a system and I couldn't use the methods that I tell people over and over again. So I do feel like it was a major thing that I learned. . . . It was worth the frustration to get what I think of as a lasting understanding of place value. (Schifter & Fosnot, 1993, p. 60)

● *Multidigit calculation provides opportunities to both deepen understanding of place value and build meanings for operations.*

The following errors are commonly seen in elementary classrooms:

$$
\begin{array}{ccc}
26 & 43 & 54 \\
+58 & -29 & \times 23 \\
\hline
714 & \overline{26} & \overline{162} \\
 & & 108 \\
 & & \overline{270}
\end{array}
$$

In these examples, children are applying their single-digit math facts but are mis-remembering their computational procedures. Because they are not thinking about the size of the numbers they start with or about what the operations do, they form no reasonable estimate of the outcome. If neither the children, nor their teachers, have learned to approach such problems with the expectation that they should make sense, it is difficult to correct the misconceptions underlying these errors.

Teachers with richly developed meanings for the operations (a sense of the variety of situations and representations associated with the operations) and a flexible understanding of place value (for example, knowing how to decompose numbers into convenient parts and operate on them) are positioned to help such children. They can recognize the strong thinking of children like Gabriella, Roberto, and their classmates in Scene 1, and they can help children who make such errors as those shown above go back to what they *do* understand about numbers and operations in order to help them recognize their errors.

Solving multidigit problems in their heads—"mental arithmetic"—and then sharing the strategies they employ, is an especially useful exercise. Teacher educators have found (Schifter et al., 1999b) that many students come to courses believing that conventional algorithms offer the only valid methods of computation. Those who invent their own strategies often feel sheepish, as if they are relying on "crutches," or are embarrassed by their lack of "sophistication" Once the hold of these prejudices is loosened and they begin to maneuver about the number system more fluently, they begin to see how the base-ten structure can be used flexibly and efficiently.

As various methods of calculation are encountered, teachers must consider the logic behind each: Does this method always work? Some need to consider why very basic procedures are justified, say, $58 + 24$: $50 + 20 = 70$; $8 + 4 = 12$; $70 + 12 = 82$. In one seminar, a teacher watching a video of second graders solving problems such as this, blurted out, "I can imitate this method to apply to other numbers, but I don't see why it works. It's just another meaningless algorithm to me!" In this situation, representations such as blocks or number lines, which help teachers think about what the operations *do*, can help them see a justification for the procedures. Some teachers must think through the general principle that addends can be decomposed and the parts recombine in any order, yet conserve the sum; or that when subtracting, if the same amount is added to or subtracted from both quantities, the difference remains constant.

Other, more complicated procedures are often challenging to teachers. Consider the following steps, commonly devised by primary grade children, for solving $35 - 16$:

$$30 - 10 \text{ is } 20$$
$$5 - 6 \text{ is "1 in the hole"}$$
$$20 - 1 = 19.$$

This procedure raises such questions for teachers as: Why is the 1 subtracted rather than added? Will this work for any subtraction problem, even one with numbers larger than two digits? When does this method apply and what comparable method can be used when it doesn't? Can the steps of this procedure be articulated as an algorithm? A college student writing in her journal, excerpted below, touches on these issues.

> I have been amazed at how this "thing" we call place value has come to make real sense to me. This goes beyond the traditional breakdown of a number. For example, I know the number 84 is comprised of 8 tens and 4 ones, but the way I look at doing a math problem is beginning to change. For example, when I look at the problem $84 + 76$, I can now do it several different ways. I can look at $84 + 76$ and say to myself:
>
> $80 + 70 = 150$ and $6 + 4 = 10$; add $150 + 10 = 160$; or
> $84 + 70 = 154$ and $154 + 6 = 160$; or I can revert to my
> traditional method:
>
> $$\begin{array}{r} 84 \\ +76 \\ \hline 160 \end{array}$$
>
> Furthermore, I am able to apply this same type of thinking to subtraction problems. I have more difficulties with subtraction problems, but I am working on increasing my comfort level. What I discovered to be very interesting were the many ways a subtraction problem could be broken down. . . . The example of $35 - 16$ was a great one. This led to many different discussions. I was able to look at this problem and say:
>
> $5 - 6 = -1$, $30 - 10 = 20$, and $20 - 1 = 19$.
>
> However, a fellow student made it even clearer by lining up the problem in a more systematic way:

$$\begin{array}{r|r} 3 & 5 \\ -1 & 6 \\ \hline 20 & -1 \end{array} = 19$$

(Student journal, spring, 1999)

This aspiring teacher, her classmates, and others enrolled in comparable classes certainly know the computation algorithms before beginning the course. What they learn is flexibility in decomposing numbers, figuring out how to recombine them to perform the operations, and thinking about the operations in terms of the actions they model. The numbers they are operating on remain in view and do not get lost in a thicket of disconnected digits. (Thus, "$3 - 1$"; in the computation above is correctly recognized as a representation of $30 - 10$, which equals 20.)

Similarly, with multidigit multiplication and division, learners (both children and teachers) first think in terms of groups in order to sort out calculation procedures. Again, teachers can begin an exploration of multiplication through practicing mental calculations. Then they can analyze their own, their classmates', and children's methods of calculation. Once the idea of rectangular array is introduced into thinking about groups, the area model, in particular, brings to light the partial products of two-digit multiplication. Later, these ideas can be formalized as the distributive property.

- *Comparing procedures can make the reasoning behind algorithms transparent*

In Scene 1, Ms. Lo Cicero asks the class how Roberto's and María's methods are alike. In this way, she highlights particular steps in the procedures and draws students' attention to analogous lines of reasoning in the different representations. For another example of comparing procedures, consider the following scene.

> **Scene 3, from a fifth grade classroom:** The class has been given the homework problem $728 \div 34$. One child, Henry, presents this solution method:

$$34 \times 10 = 340$$
$$34 \times 20 = 680$$

$$\begin{array}{r} 680 \\ + \ 34 \\ \hline 714 \end{array} \qquad \begin{array}{r} 728 \\ -714 \\ \hline 14 \end{array}$$

> Henry explains to the class, "Twenty 34s plus one more is 21. I knew I was pretty close. I didn't think I could add any more 34s, so I subtracted 714 from 728 and got 14. Then I had 21 remainder 14."

> Another child, Michaela, presents her solution:

$$\begin{array}{r} 21 \\ 34\overline{\smash)728} \\ 68 \\ \hline 48 \\ 34 \\ \hline 14 \end{array}$$

> Michaela describes the steps of the conventional division algorithm: "34 goes into 72 two times and that's 68. You gotta minus that, bring down the 8, then 34 goes into 48 one time."

> Apparently, their teacher has not shown the conventional division algorithm to her students, and Michaela's classmates say they don't

understand her solution. Asked to explain, Michaela takes the class through the steps again, but with the same response. Then the teacher asks the class to compare the two procedures to identify similar parts, assisting them by inserting a "0" next to Michaela's "68" so that the children could more easily see where Henry's 680 shows up in Michaela's process. Through the discussion that ensues, using Henry's solution as a point of reference, some of Michaela's classmates can begin to see the justification for the steps she had taken.

(based on NCTM, 2000, pp. 153–154)

The reasoning behind Henry's method is clear to him and his classmates. But when Michaela presents the conventional long division algorithm with its more efficient notation, the rationale for her procedure eludes both her and her classmates. The teacher recognizes the parallel reasoning behind the two methods and draws her students' attention to it, thus giving them access to what was an initially opaque process.

The steps of the conventional algorithms, particularly for multidigit multiplication and division, are often every bit as mystifying to teachers as they are to children. The former, too, can compare procedures, devised by themselves or by students, or by other cultures, to bring to light their conceptual bases.

It is also useful to examine commonly applied *in*correct procedures for solving multidigit multiplication problems, such as those instantiated in the following strategies observed in the work of teachers and children.

Teachers or children calculated 16×28 by:	Writing:
Operating on the tens, operating on the ones, and adding the results.	$(10 \times 20) + (6 \times 8) = 248$
Subtracting 2 from one factor and adding it to the other; then operating.	$14 \times 30 = 420$
Rounding up to the nearest tens, operating, and subtracting off what had been added on.	$(20 \times 20) - 4 - 2 = 594$

Each of these incorrect methods derives from misapplying additive procedures. After all, when *adding* $16 + 28$, one can operate on the tens, operate on the ones, and add the results, etc. By analyzing these procedures, teachers have opportunities to deepen their understanding of multiplication and the distributive property, and to become sensitive to the tendency to extend additive procedures to multiplicative situations.

Presenting and exploring these various methods of calculation highlights for teachers the differences among the operations. In addition and subtraction, the units are the same; in multiplication and division, more than one unit is involved. In addition or subtraction, one can decompose the addends or both the minuend and subtrahend, respectively. In multiplication, additively decomposing both factors frequently makes the calculation more complex. And in division, additively decomposing the divisor is not useful. Up to now, this discussion has been confined to the arithmetic of whole numbers. And children do begin to learn about numbers through counting. Soon, though, the world of number expands to include integers and rationals. In order to support children through this transition, teachers, too, must have explored these new concepts.

• *As with whole numbers, teachers must learn to give meaning to operations with integers.*

Many young children are exposed to numbers less than 0 outside of school, through discussions of weather in the wintertime, say, or by keeping track of scores in some of their games. In an example cited above, a second grader says that $5 - 6$ equals "one in the hole" and knows how to use that idea to compute $35 - 16$. However, in general, operating with integers presents new issues.

> **Scene 4, from a third grade classroom:** The children have been working with an image of an elevator to represent integers. The ground floor is 0; floors are numbered up to 12 to the roof and to -12 below the ground level. The children write number sentences to model "elevator trips." For example, if a person starts on the third floor and goes down seven floors, the trip is represented as $3 - 7 = -4$. The children can do this task well and come up with significant observations:
>
> Nathan: Any number below zero plus that same number above zero equals zero.
> Ofala: Any number take away double that number would equal that same number only below zero.
>
> However, the teacher is concerned about the limitations of the elevator representation. For example, it allows the children to think about subtracting a positive integer as "going down" or about subtraction as the distance between floors, but the representation does not help the children develop a sense of "taking away" numbers less than zero. Nor could they make sense of certain addition expressions, e.g. $6 + (-6)$.   (based on Ball, 1993)

Knowing the rules for computing with integers is insufficient for understanding operations with numbers less than zero. As with whole-number operations, teachers and children must learn to think of the variety of situations that can be modeled by the operations. Now, however, as the numbers represent both magnitude and direction, the situations increase in complexity.

• *Fractions introduce a new kind of number.*

Children are introduced to rational numbers through their work with fractions. Although most young children are familiar with the numbers $1/2$, $1/4$, and perhaps $3/4$, the idea of fraction is challenging. To many, these numbers represent a quantity less than one, or, perhaps, part of a whole, but they might also talk about how "your half is bigger than my half" or be unable to interpret the meaning of, say, $2/3$. And even when children seem to understand the meaning of fraction in some situations, that understanding often proves fragile and context dependent. For example, in a class of third graders who had been working on fractions for some weeks, the question arose, Which is larger, $4/4$ or $5/5$? Some argued for $4/4$ because the parts are larger; others, for $5/5$, because there are more parts. No one argued they were equal (Ball & Wilson, 1996).

Adults are generally unable to recall a time when their concept of number was exhaustively defined by the experience of counting whole numbers. Yet, listening to children being introduced to the idea of fraction and realizing how this challenges their very notion of number, offers adults an opportunity to think through how the concept of number expands as one moves from the system of integers to rationals. It is no longer merely a matter of counting units. Instead, one must now count the number of units in one quantity, count the number of units in a second quantity, and derive a third number—a new *kind* of number—that places the first quantity *in relation* to the second (Behr & Hiebert, 1988; Carpenter, Fennema, & Romberg, 1993).

Many children, and older students as well, see fractions only as pairs of natural numbers plugged into arithmetic procedures. So, for example, in the second National Assessment of Educational Progress, when students were asked to pick an estimate for $12/13 + 7/8$ from the choices, 1, 2, 19, and 21, most chose the latter two, presumably having combined either their numerators or their denominators. They failed to recognize that $12/13$ and $7/8$ are each quantities close to 1 and, thus, their sum is close to 2 (Carpenter et al., 1981).

For teachers to be able to perceive the mathematical ideas children must put together in order to develop the idea of fraction, their own understanding of the concept must be expanded. The same number created by finding the part of a whole can also be seen as an expression of division (further discussed below), as a point on the number line, as a rate, or as an operator.

Most teachers know what a fraction is under at least one of its interpretations, but they often lack a sense of relative size. Having memorized a method for finding a common denominator and comparing numerators, they cannot determine, say, which of a pair is larger—$5/7$ or $7/9$? $5/8$ or $7/12$?—without applying that procedure. Working with area diagrams, teachers can explore such fraction pairs and learn to use other strategies, e.g., given common numerators, comparing the denominators; or considering how much smaller each fraction is than 1, or how much larger than $1/2$. Such observations can, in turn, be applied to more cumbersome pairs of fractions.

• *Developing meaning for calculating with fractions enriches understanding of both fractions and operations.*

The following vignette illustrates two major mathematical ideas—unit and fraction as quotient—that need to be investigated by teachers and children, too.

> **Scene 5, from a fifth grade classroom:** To work on a representation of "3 divided by 15," children produce the scenario "three pizzas need to be shared among 15 people," and solve the problem by dividing each of the pizzas into five slices. When the teacher changes the problem to 3 divided by 16, there is not such a neat solution.
>
> One child suggests cutting each circle into 16 pieces. Another, whose scenario is "share 3 pieces of cheese among 16 mice," says that if you divide each piece into five slices, one mouse won't get any. Another girl says she should have divided each of the three pieces of cheese into six slices. A debate ensues about what to do with the extra slices. Some of the students think it would work to divide each of the leftover slices into eight pieces, and they are challenged to explain to their classmates how they have figured out that it will be eight. Another issue for debate is what each of the little pieces should be

called: one sixteenth? one eighth? one forty-eighth? (Each of the slices, it has been agreed, is one sixth of a piece of cheese.)

(based on Lampert & Ball, 1998, pp. 139–141)

The first major mathematical idea, "unit," is central to work with fractions. In Scene 5, the children correctly suggest that the small bits of cheese could be represented as $1/16$, $1/8$, or $1/48$—depending on the unit chosen. The second major idea is the relationship between fraction and division. Many teachers have considered fractions only as parts of a single-unit whole—e.g., $3/16$ involves dividing a single whole into 16 equal pieces and taking three of those. The idea that $3/16$ might mean three wholes divided equally into 16 portions is new to them. And they would have difficulty explaining how the quantity $3/16$ is the same as the quantity $1/6+1/48$ (Lampert & Ball, 1998, p. 142).

In the Mathematics and Teaching through Hypermedia Project, prospective teachers work on these ideas, analyzing videotape of the fifth graders' discussion summarized in Scene 5, and supplemented by problems the teachers solve for themselves. For example, they are given the following assignment:

- Think about the following interpretations of 3 divided by 17.

  3 pizzas divided among 17 people
  3 dollars divided among 17 people
  3 dozen donuts divided among 17 people

- Write or draw an explanation of how you might do each of these "fair share" problems.

- Now try dividing the same quantities of pizza, money, and donuts among 15 people. What different math gets called into play?

- Now try dividing the same quantities of pizza, money, and donuts between 2 people. What different math gets called into play this time? (p. 143)

This assignment requires teachers to work with different kinds of units, prompting them to investigate different representations and the computations associated with them. Changing the number of people involved in the problem challenges the teachers to consider what it is about the numbers that makes the problem come out as it does; teachers then make conjectures about what would happen with other numbers. The issue of unit arises again when adding fractions. Consider the problem, *One batch of muffins needed 3/4 cup of flour. The second batch needed 2/4 cup of flour. How much flour was used in both batches?* In one class of fourth graders working on this problem, some children argued for the answer $5/4$, others $5/8$. All were looking at the following representation (Heaton, 2000).

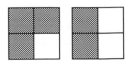

In order to help her students work out why the correct answer is $5/4$, first the teacher needed to see why $5/8$ made sense to some of her students, what would be a question in this context whose answer is $5/8$ (e.g., If you start with 2 cups of flour, how much of the flour do you need for the two batches?). Understanding how $5/8$ could make sense to some of her students, she was in a better position to help them see why $5/4$ is the sum of $3/4 + 2/4$.

Addition and subtraction require working with single units, but multiplication and division involve more than one. This is precisely what is so difficult about devising word problems or diagrams for, say, $3/4 \div 1/2$: What is the unit for $3/4$? for $1/2$? for the quotient? And how does that shift when multiplying $3/4 \times 1/2$? On the other hand, sorting out these issues can bring to light the reasoning, so elusive to both children and adults, behind the invert-and-multiply algorithm. (These ideas are further elaborated in the chapter on mathematics for middle grades teachers.)

• *Decimal fractions extend the ideas of place value to numbers less than 1; as with calculations with common fractions, decimal computation can enrich understanding of the operations.*

> **Scene 6, from a fifth grade classroom:** In September, the teacher had given the class a set of word problems, among them, "Rob wants to read one hundred pages of his book before his next conference in seven days. How many pages should he read each day?" Now, two months later, she asks her students to look at the problem again, but to find the answer on a calculator. The children all report, 14.285714. After reviewing an earlier discussion about the interpretation of their original answer, 14 remainder 2, the teacher asks "What is the '.285714' in the calculator's answer?"
>
> The class begins to talk about this and, after a few minutes, Jeremy raises his hand. "I think I get it. In the 14.285714 it's like the 2 is a paragraph and the 8 is a sentence and the 5 is a word and the 7 is a letter and the 1 is part of a letter. I don't know what the 4 is. Only the 14 [the two left-most digits] really counts anyway. The other pieces are really small, especially after you get beyond the sentences." "That's pretty interesting," the teacher says. But recognizing the limitations of Jeremy's metaphor, she offers him an opportunity to qualify it: "Does it make sense to you that it could work that way?"
>
> "Well, it doesn't really make sense," he answers. "I mean, you don't have pieces of words to be read and things like that. It does make sense in some ways though; like how I said, it's really only the first few numbers that make a difference. The rest are too small to matter."   (based on Schifter et al., 1999c, pp. 108–110)

Many children and teachers, too, believe decimals bring with them a new set of rules to remember, but the main principles that underlie decimal fractions are the same as those that govern whole numbers. Precisely because they are tacit with respect to whole numbers, formerly unproblematic concepts now need to be considered, thus providing an opportunity to return to whole numbers with new insights.

For example, in Scene 6, a fifth grade class is presented with an eight-digit number between 14 and 15 and the children are challenged to interpret the digits to the right of the decimal point. Although Jeremy's metaphor is limited (a paragraph is not necessarily $2/10$ of a page; a sentence is rarely $8/10$ of a paragraph), it does bring him and his classmates to an important observation: "It's really only the first few numbers that make a difference. The rest are too small to matter." Indeed, this can be said of any multidigit number, whole ones included. The digits to the left are more significant than those to the right in determining the magnitude of the number. (This is the basic idea behind scientific notation; and the same principle is highlighted in stem-and-leaf representations of data.)

Like the fifth graders in Scene 6, teachers, too, need to develop meanings for the digits to the right of the decimal point. Having them create their own representations of decimals (with blocks or diagrams) provides opportunities to bring out misunderstandings and highlight issues that need to be explicit for teachers— what is the value of each place and how do the digits combine to represent a single quantity?

Teachers must come to see that any number can be approximated arbitrarily closely by finite decimals. The study of repeating decimals invites work with the calculator and, especially given the calculator's finite capacity, exploration of its limitations. Why is it that a calculator cannot exactly represent $1/3$? How can you characterize all fractions that have finite representations?

To summarize this discussion of decimals, consider the following journal excerpt, by an inservice instructor who asked her class to compare whole number and decimal addition.

> People at first posited many differences, but by the end of the discussion came to see none! For example, the first thing stated as a difference was that you line whole numbers up from right to left and then you line decimals up from the decimal point. Then someone suggested that if you have two decimals with the same number of places, you also line them up from the right and that, in fact, you are always (no matter what kind of numbers you are dealing with) lining up like places with like places and that this was a similarity, not a difference. . . .
>
> Another difference mentioned was that when you add decimals, the quantities got smaller, but when you add whole numbers, the quantities got bigger. People actually thought about this for a moment. I suggested we add some decimals, so they proposed, $.15 + .16$. Everyone then stated that no, in fact, even decimal addition makes numbers bigger [when only positive addends are under consideration]. Someone stated authoritatively that when you multiply, the result ends up being smaller. I asked disingenuously how this could be since just last session people were telling me that multiplication was nothing more nor less than repeated addition. If addition of decimals produces larger quantities, how could multiplication of decimals produce smaller quantities? Some people laughed. . . .
>
> The third difference was proposed by Nancy, who said that the places got smaller as you go to the right of the decimal and larger as you go to the left. I asked her to come up and show us what she meant and she approached the board, pulled up

sharply, and said, "Oops, never mind!" as she figured out that even to the right of the decimal, the farther to the left you go, the bigger the number.

Then someone suggested that regrouping changes as you move across the decimal. I wondered how, and several people said it worked the same way regrouping works in whole number addition, but that the places had different names.

(Yaffee, personal communication, 1997)

This recap of a discussion among teachers working to understand number and operations highlights several themes. First is the habit of noticing superficial characteristics of calculation procedures. When dealing with rule making at this level, one must remember different rules for the different kinds of numbers. Moreover, when relying solely on memory, one is likely to come up with such misremembered "facts" as "adding decimals produces a sum smaller than the addends," but with no resources to challenge them. Returning to basic principles, however, different rules may merge into one. The arithmetic for decimals is essentially the same as for whole numbers.

Second, shallow or mistaken ideas, such as those offered in this recap, can sit alongside correct understandings. When the classroom environment is safe enough to bring such notions out into the open, they can be challenged, corrected, replaced, or modified.

And third, this journal entry illustrates how the basic themes of place value and operations recur in the context of work with rational numbers.

One might have noticed that, in this discussion of number and operations, the formal statements of properties (some prefer to call them "the laws of algebra") have not played a central role. To many teachers at the elementary level, algebraic notation obscures rather than illuminates. Introducing formal axioms in the expectation that teachers will be struck by the beauty and logical economy of our number system is naïve. Only after they have done the kind of work described above—have come to know the various kinds of situations modeled by the operations, developed a variety of representations for them, and worked with these representations to explore calculation with whole numbers, integers, and rationals—can they make sense of algebraic notation. Without it, mechanical application of rules (commutativity, associativity, etc.) is likely to leave them with those familiar feelings of disconnection.

Work with algebra is discussed in the following section.

Algebra and Functions

Although the study of algebra and functions generally begins at the upper-middle- or high-school grades, some core concepts and practices are accessible at a much earlier age. If teachers are to cultivate the development of these ideas in the elementary grades, they must understand those concepts and practices and recognize how they are manifested in the mathematical thinking of young children.

- Generalizing arithmetic and quantitative reasoning:

  - learning to use a variety of representations, including conventional algebraic notation, to articulate and justify generalizations.

- understanding algebraic expressions as shorthand for describing calculation; understanding algebraic identities as statements of equivalence of expressions.
- understanding different forms of argument and learning to devise deductive arguments.
- solving word problems via algebraic manipulation.

- Discovering how the field axioms govern arithmetic:

  - recognizing commutativity, associativity, distributivity, identities, and inverses as properties of operations on a given set.
  - seeing computation algorithms as applications of particular axioms.

- Understanding functions:

  - becoming familiar with the notion of function.
  - being able to read and create graphs of functions, formulas (closed and recursive), and tables.
  - studying the characteristics of particular classes of functions on integers, especially linear, quadratic, and exponential functions.

When children begin their study of algebra in middle or high school, they learn a new language, an efficient way of representing properties of operations and relationships among them. Now they are expected to make meaning for such sentences as,

$$\text{If } a > b \text{ and } c > d, \text{ then } a + c > b + d$$
$$a^2 b^2 = (ab)^2$$
$$(x+1)^2 = x^2 + 2x + 1$$
$$(2n+1) + (2m+1) = 2(n+m+1)$$

If, in earlier grades, students lose their ability to make sense of mathematics and, as a consequence, can attach no meaning to arithmetic expressions, they have nothing on which to build their algebra. On the other hand, to those already familiar with those properties and relationships, the challenge is learning the conventional system of notation.

When the elementary classroom is designed to encourage and build upon children's thinking—where students pursue their own questions—we find them interested in formulating and testing generalizations (Ball & Bass, 2000a, 2000b; Carpenter & Levi, 1999; Russell et al., 1999). This is particularly evident in their work on calculation and number theory topics: evens and odds, square numbers, factors. Children's interest in articulating these generalizations provides an opportunity for them to explore the ideas they will later learn to express in algebraic form. At the same time, it provides an opening to work on methods of justification.

Many elementary teachers have shared the situation of the child who enters algebra class without a sound background in arithmetic. They, too, struggled through their courses, memorizing rules for manipulating symbols. If, on the other hand, prospective teachers are offered a course that helps them make sense of number and operations (as described above), then they are prepared to learn to use algebraic notation to express relationships that have meaning for them.

However, to be able to support children in the classroom, teachers will need more than fluency in algebra. They must appreciate the power of generalization, be able to recognize when children are approaching this territory, and understand what counts as a justification.

• *To build on children's capacities to articulate their observations and to generalize requires teachers who understand the importance of generalization and who command a variety of methods of justification and forms of representation.*

> **Scene 7, from a kindergarten classroom:** The children are in pairs playing a version of the card game War. For each round, they each put down two cards and whoever has the larger sum takes the four cards. Myra and Janie have just laid out their cards and Myra declares, "I get these." Janie protests, "But you didn't count yet! I might have more." Myra explains, "My two numbers are more than your two numbers, so when you put them together, mine is more."
>
> (based on Seyferth, field notes, 1995)

> **Scene 8, from a second grade classroom:** The children have become intrigued by square numbers (squares of natural numbers) and have set out to learn whatever they can from them. They work in small groups and, as each group makes a new observation, a child goes to the chart paper the teacher has set up in the front of the room and writes it down. At the end of the session, the list includes the following items:
>
> • 1, 4, and 9 are square numbers.
>
> • 16, 25, 36, 49, 64, 81, and 100 are square numbers.
>
> • If you times a square number by a square number, you get a square number.
>
> • Take any square number and add two zeros to it and you will get another square number, like 4, 400.
>
> (based on Rigolleti, unpublished paper, 1991)

> **Scene 9, from a combined third/fourth grade classroom:** The class was given a problem involving eight odd numbers that summed to 71. Now, in whole group, the children discuss how they know this is an impossible situation. The following arguments are offered:
>
> • You have to try it a bunch of times.
>
> • It goes odd, even, odd, even, odd, even [each time you add an odd number].
>
> • I know that an odd and an odd always equal an even. [In the problem,] there are eight different kinds, so each one has a partner to equal an even and the evens can't equal up to 71.
>
> (based on Bastable & Schifter, in press)

In Scene 7, Myra's strategy for determining which child has the larger sum subverts her teacher's goal for the lesson. Rather than practice counting or adding, Myra reasons about quantities in general (Thompson, 1993). A teacher who recognizes the power of her observations will sacrifice her immediate objectives in order to encourage such thinking.

Given a situation similar to that depicted in Scene 8, a teacher who registers the difference between the first and second pairs of statements can ask about the latter, will it always work? Indeed, the teacher who recognizes that the fourth statement is a corollary of the third (understanding that the children's meaning for "add two zeros to it" is "insert two zeros after the last digit") is positioned to assess whether any members of the class can take on that idea.

And the third and fourth graders in Scene 9 make inferences based on three different kinds of arguments: testing a conjecture on a set of specific numbers, reasoning by extending a pattern, and forming a deductive argument. Only teachers who themselves appreciate the differences among these different forms of justification can, in turn, help their students understand them.

Greater attention to algebraic thinking at the elementary level has encouraged teachers and researchers to look into young children's abilities to reason with variables (Ball, 1989; Carraher et al., 2000). For example, in one first and second grade combination class (Carpenter & Levi, 1999), students worked with such equations as,

$$
\begin{aligned}
x + x + x - x &= 10 \\
x + y - y &= x \\
x + x &= y.
\end{aligned}
$$

Children working on the first equation concluded that the sentence is true when $x = 5$. They generated the second when asked to find "an open number sentence that is true for every number, no matter what you put in." For the third, they found a number of solutions and recognized that y must be twice as large as x. As this kind of material moves into elementary classrooms, teachers must be aware of different uses of variables: to express unknowns that can be solved, to express identities, and to express relationships between sets of numbers (Usiskin, 1988).

To work on these issues, teachers, too, must be given contexts—perhaps, like their students, to explore factors, divisibility, square numbers—in which they can come up with their own observations and assess the validity of their own and their classmates' claims. For elementary teachers, it is especially important that they learn to develop arguments using, in addition to algebraic notation, representations familiar to them. For example, the third and fourth claims in Scene 8 can be supported using an area model of multiplication. By identifying the same lines of reasoning in strategies employing different forms of notation, teachers can now learn to give meaning to once meaningless symbols. And as they become used to algebraic notation, they can learn to appreciate its power and flexibility, especially in comparison to geometric representation.

For elementary teachers, an understanding of commutativity, associativity, distributivity, identities, and inverses cannot be taken for granted. Lacking proper grounding, many will have the sense that these rules have simply been "pulled out of the air." Rather, they should be given tasks that help them make these generalizations from their experiences. For example, one might ask teachers to consider these properties for all four operations, devising representations or situations to demonstrate why they work when they apply, why they don't if they don't, and then to identify any patterns they discover. Although it may be obvious to teachers that addition and multiplication are commutative, it is not obvious that reversing the quantities when subtracting or dividing yields the inverse. Although associativity of addition may be clear, teachers need to think through why, for example, $(10 - 7) - 2$ results in a smaller number than $10 - (7 - 2)$.

In one inservice class, teachers exploring these properties reported the following conclusion: "When you reverse the numbers for subtraction, you get the opposite— except when the numbers are the same." To understand what the teachers were saying, the instructor worked with them to represent their observation in conventional algebraic notation: If $a - b = c$, then $b - a = -c$, except when $a = b$. Perplexed by this exception, the instructor challenged them. After a few minutes of confused discussion, the teachers explained that if $a = b$, then the answer is 0 both ways, not -0, which they called "the opposite of 0." When asked for the opposite of 0, there was a long pause, until someone ventured, "Infinity—the opposite of nothing is everything." This opening provided an opportunity to think about additive and multiplicative inverses, additive and multiplicative identities, and how definition plays a role in creating a consistent system (Schifter, field notes, 1989).

With a deeper understanding of commutativity, associativity, distributivity, identities, and inverses, teachers can return to calculation algorithms—those conventionally taught in the United States, those taught in other countries, as well as those children frequently devise—to analyze them as applications of these properties.

• *Solving multistep problems in a variety of ways also provides opportunities for teachers to create meaning for algebraic notation.*

Problems conventionally solved using algebra are, in fact, accessible to elementary grade children, as well as to teachers whose algebraic skills are under-developed. For example, consider the following problem (written quite a few years ago, when we paid less for our fruit): Ten apples and 5 bananas cost $1.65, and one apple and one banana together cost $.20. What is the cost of one apple? one banana? In one course for teachers, some teams got started by pulling out colored cubes: 10 red for apples, 5 yellow for bananas. Each cube stood for the cost of a single piece of fruit and, together, the 15 cubes represented $1.65. Pairing red and yellow cubes, then discarding them, the teachers subtracted $.20 for each pair. Left with 5 red cubes worth $.65, they concluded that each was worth $.13. Therefore, a banana cost $.07.

Having solved the problem in a way that called upon their own mathematical ideas, the teachers could then represent it algebraically. Identifying a as the cost of an apple and b as the cost of a banana, they could now see parallels between the steps they took with the blocks and the steps of the algebraic procedure.

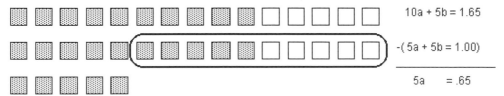

$$10a + 5b = 1.65$$
$$-(5a + 5b = 1.00)$$
$$5a \quad = .65$$

As teachers come to see how algebraic representations correspond to representations of actions modeled by other solution methods, they become more confident and skillful with algebra. From here, they can appreciate the use of algebra in problems for which other representations are too cumbersome.

- *Work with patterns has been a part of the K–4 curriculum, but the concept of function is new.*

> **Scene 10, from a kindergarten classroom:** The class has been working on patterns. As the teacher presents today's pattern (green, orange, brown, green, . . .), she also writes down the number associated with each element: "1" is written under the green square, "2" under the orange, "3" under brown, "4" under green. As has been routine, the children call out what comes next: orange, brown, green. Now the teacher points to the numbers she has written to show how she has identified the place of each square and poses a new question to the class: "What color square will go in the tenth place?" Different children call out different colors, as if this were a guessing game. But Roberto speaks forcefully. He stands up, stamps his foot, and declares, "It's green! I know it's green!"
>
> (based on Cohen, personal communication, 1994)

> **Scene 11, from a third grade classroom:** The teacher is working with a small group of children who are interested in the surface area of towers they make out of cubes. Starting with a $1 \times 1 \times 2$ tower, it takes some time for the children to sort out what surface area is, but eventually they conclude that 10 faces of the unit cubes show on the surface. They backtrack to look at a $1 \times 1 \times 1$ tower before they move on to a $1 \times 1 \times 3$. The teacher suggests that they organize their findings into a chart.

Number of cubes	Surface Area
1	6
2	10
3	14

> She now asks the children to think about the surface area of the next larger tower.

> Moira: How many do you add? You add 4 to each one. So the next one would be 18.
>
> Teacher: Why are you adding 4? Do you agree that every time you add on a cube, you add on 4 to the surface area? Can you explain why? Does that make sense?

> Jeff holds a stack of five cubes. He then shows the group how he counts all the square units on each face of the tower—$5+5+5+5+2$. He points to the 5 square units on each of the four vertical faces, then adds the 2 for the ends. But then he shows that it is also $4+4+4+4+4+2$ if he counts around all the vertical square units of each cube in the tower and then adds the ends. He says it's $(4 \times 5) + 2$ and $(5 \times 4) + 2$. The others agree.

> Two days later, the teacher gives the group cubes of different sizes—with edges of 1, 2, 3, and 4—and asks what they think the surface area of each is. The children record their findings: 6, 24, 54, 96—and Jeff declares the next one is 150, though he hasn't made one that size.

> The teacher asks if the group could make any rules to describe how the areas change. First Jeff says the areas were all 3—numbers (multiples of 3). Ron says they are also counted by 6—numbers. Then he says the areas were always six times one face of the whole cube. Ron suggests that they added three 6s every time. He points out the

distance between 6 and 24 is 18 and guesses the same is true for 24 to 54. When he sees the distance is 30, Jeff points out that's five 6s. They figure out it's seven 6s between 54 and 96 and observe: 3, 5, 7. So Jeff says it must be nine 6s to 150.

(based on Schifter et al., in press)

Elementary teachers often create rich classroom experiences around patterns, usually beginning with a sequence of elements (sometimes in the form of sounds or movements), where a set of elements forms a unit that is repeated. Children learn to extend the pattern by determining and repeating the unit. Later, children have experiences with number patterns—finding the next number, then explaining how they did it. However, teachers rarely have a sense of how this is related to the mathematics their students will encounter in later years: specifically, that a function can be created by labeling the elements or units of a pattern by the natural numbers. For example, the kindergartners in Scene 10 have been presented with a function, and based on the regularity of the pattern, Roberto is able to identify the value in the range associated with a particular value in the domain.

Some of the patterns children find particularly intriguing point toward ideas that will prove significant in higher grades. For example, second graders working on square numbers notice that as they increase, a pattern forms. The numbers go up by 1, 3, 5, 7 . . . Third graders in Scene 11 notice a similar pattern for the surface area of cubes: As the length of each side increases by 1, the surface area increases by 3 sixes, 5 sixes, 7 sixes.

Those same third graders also notice a different pattern when they count the square faces on the surface of the $1 \times 1 \times n$ towers they build: The numbers go up by 4. Teachers who are familiar with the characteristics of particular classes of functions—in this case, linear and quadratic—would have a context for their students' observations.

The children's observation that "you add 4 to each one" is consistent with a recursive representation of functions: $f(n+1) = f(n) + 4$. Jeff points out how you can look at a 5-cube tower and see the surface area as $4 \times 5 + 2$, approaching a representation in closed form: $f(n) = 4n + 2$. Teachers must be familiar with both.

Some of the new curricular materials for the elementary grades have children working with stories, graphs, and tables to describe situations that change over time, e.g., height, speed, distance, population. These include graphs not defined by elementary functions. The very ideas on which the children are working are content for teachers: What is the meaning of a horizontal straight line? a line tilted up? a line tilted down? What is the relationship between rate of change and accumulated change?

Those familiar with the usual college-level algebra course will recognize that it does not address the issues described in this section.

Geometry and Measurement

For many years, the geometry curriculum for the elementary grades mandated recognition of basic two-dimensional shapes, measurement of length with standard and non-standard units, and the ability to apply area and perimeter formulas for rectangles (and possibly a few other shapes). Because so many students entered high-school geometry courses unprepared for them (Usiskin, 1987; van Hiele, 1986), topics in geometry have recently been given a more prominent role in the early grades. Their own experience with high-school geometry notwithstanding, for most elementary teachers, much of this material—some of it highlighted below—is new.

Summary of geometry and measurement content.

- Developing visualization skills:

  - becoming familiar with projections, cross-sections, and decompositions of common two- and three-dimensional figures.
  - representing three-dimensional shapes in two dimensions and constructing three-dimensional objects from two-dimensional representations.

- Developing familiarity with basic shapes and their properties:

  - knowing fundamental objects of geometry.
  - developing an understanding of angles and how they are measured.
  - becoming familiar with plane isometries—reflections (flips), rotations (turns), and translations (slides)—and symmetries.
  - understanding congruence and similarity.
  - learning technical vocabulary and understanding the importance of definition.

- Understanding the process of measurement:

  - recognizing different aspects of size.
  - understanding the idea of unit and the need to select a unit appropriate to the attribute being measured.
  - knowing the standard (English and metric) systems of units.
  - comparing units.
  - understanding that measurements are approximate and that different units affect precision.

- Understanding length, area, and volume:

  - knowing what is meant by one, two, and three dimensions.
  - seeing rectangles as arrays of squares, rectangular solids as arrays of cubes.
  - recognizing the behavior of measure (length, area, and volume) under uniform dilations.
  - devising area formulas for triangles, parallelograms, and trapezoids; knowing the formula for the area of a circle; becoming familiar with formulas for prisms, cylinders, and other three-dimensional objects.
  - understanding the independence of perimeter and area; surface area and volume.

Spatial visualization—building and manipulating mental representations of two- and three-dimensional objects and perceiving objects from different perspectives—is a critical aspect of geometric reasoning (Battista, 1999; National Council of Teachers of Mathematics, 2000), but is a capacity few teachers have had opportunities to develop. Helping teachers cultivate spatial sense must be an initial, and is certainly an indispensable, goal in a geometry course.

• *Developing spatial sense and identifying significant features of shape are the core tasks of elementary-level geometry and the basis of what teachers have yet to learn.*

The activities illustrated in the following scenes drawn from fourth grade classrooms prove equally stimulating and challenging for teachers.

Scene 12, from a fourth grade classroom: A three-dimensional image is shown on a screen for three seconds, and the children are asked to build a copy of it with interlocking cubes. They go through this process with a set of structures, and for each one, stop to discuss how they saw the structure (Battista & Clements, 1998). The image below takes the most time to discuss because some students feel strongly that this image can only have nine cubes, but others think it could have ten.

Natalie explains that she saw three first, with two added to the top end, two more going down, and two behind these. She explains she could start with the first three and continue from that point just adding two at a time.

Sandy agrees that Natalie's shape could be correct, but says she put one on the back bottom of the first three. She feels that one could be there and no one would see it. Robby is very interested in Sandy's idea and, using her cube building, tries putting it at the same angle as on the transparency. He wants to determine whether, looking at it from this angle, the back cube could be seen.

(based on Schifter et al., in press-a)

Scene 13, from a fourth grade classroom: The class has been given a set of non-regular polygons with the following directions: "If this were a cake you had to share evenly between two people, how could you cut it? You need to be able to explain how you know that each person would get the same amount of cake." Kumiko and Sarah have both divided the hexagon shown below by separating the triangle from the square.

The two children have different ways to explain how they know the triangle and square are equal in area.

Kumiko: I put the square on top of the triangle. Then I
          cut off the extra pieces [marked a] and put them
          in here [marked b].

Sarah: I cut this part off [the triangle marked a] and
          moved it up to the top [the triangular space
          marked b].

Another child, working on the decagon below, divides it as shown.

He explains that he can take the part on the right, flip it, rotate it,
and slide it, and it will fit exactly on top of the other.

(based on Tierney et al., 1998)

As do their students, so do teachers need experiences with activities, like those
described in Scenes 12 and 13, that cultivate spatial sense. In Scene 12, learners
construct a mental image of a three-dimensional object by analyzing it into parts,
remembering the size, shape, and orientation of each, and then recombining them
in correct relationship to one another. In Scene 13, learners decompose non-regular
polygonal shapes into more familiar ones in order to demonstrate equal areas.

Additional activities should include exercises involving projections, cross-sections, and decompositions of common plane figures—regular polygons and (as
shown here) non-regular polygons—and solids—prisms, cylinders, cones, pyramids,
spheres. (Electronic tools that support exploration, such as dynamic geometry
software, can enhance this work.)

Another set of activities for both children and teachers involves representing
three-dimensional objects in two dimensions using a variety of methods and, given a
two-dimensional representation of a three-dimensional structure, reconstructing the
original. Each method of representing three dimensions on paper preserves some
features of the original object and distorts others. Creating such representations
requires that one identify and isolate those features. Learning to read them means
being able to construct mental images of a three-dimensional object from a specific
set of cues. The children in Scene 12 are able to read the figure shown and recognize
that it is an ambiguous representation of a three-dimensional object—that two
possible structures fit the constraints of the problem.

Through this work, teachers can become familiar with basic two- and three-
dimensional shapes: identify them, be able to draw them, know their definitions
and how the shapes satisfy those definitions, recognize these shapes as parts of more
complex configurations, and know some facts about them. And through this work,
they can also develop different images of how shapes are composed. For example,
a cube can be seen as a stack of congruent squares or as an object whose surface

unfolds into a net of six squares. A tetrahedron can be seen as a stack of triangles decreasing in size or as an object whose surface unfolds into a net of four triangles.

Teachers should be able to recognize the symmetries of different polygons and solids and be familiar with some plane isometries—reflections (flips), rotations (turns), and translations (slides). As illustrated in Scene 13, children and teachers should be able to demonstrate congruence by using these isometries.

Although most teachers are able to identify right, acute, and obtuse angles, there is often serious confusion about exactly what angles are and what their measures mean. In mathematics courses for teachers, one hears such questions as: "How is it that both a large protractor and a small protractor agree on the measure of a given angle?" or "How can it be that a hexagon has 720 degrees when it fits inside a circle and a circle has only 360 degrees?" Questions like these provide an opening to discuss just what an angle is and how it is measured.

Teachers should understand the idea of angle, both as the figure formed by two rays and as angular motion. They should understand that angles can be added, that the measure of the sum of angles is the sum of the angle measures (modulo 2π or 360 degrees), and that the measure of angles of a triangle sum to 180 degrees, a straight angle. And they ought to be able to devise a proof for the sum of the angle measures of a polygon.

The following is frequently observed in professional development settings: Teachers asked to measure the interior angles of a variety of polygons soon recognize a pattern—that the sum of the measures increases by 180 degrees each time you increase the number of sides of the polygon by 1—which they can write as a formula, $(n-2) \times 180$. But then one will hear that, "When you go from a quadrilateral to a pentagon, you add two lines, but you also take away a line. Since a line is 180 degrees, you add two but take one away, so you add 180 degrees." The argument is based on the idea that "a line has 180 degrees," but takes no notice of the angles, themselves.

Similarly, teachers who present the following diagram may claim that "Any polygon with n sides can be divided into $n-2$ triangles. A triangle has 180 degrees. So the polygon has $(n-2) \times 180$ degrees."

This argument sounds like the most familiar proof of this finding, but it may be that the teachers are actually paying attention to *regions* and not to *angles*. Thus, those who make this argument are often stymied when presented with this figure:

By a similar argument, the same polygon now has an angle sum of $n \times 180$. Confronted with this contradiction, teachers must now attend to which angles of the triangles contribute to the angle sum of the polygon and express this in their arguments. The second figure illustrates another formula—$n \times 180 - 360$—which teachers should recognize as a variation of the original.

The study of geometric shapes encourages the development of technical vocabulary, establishing the power of precise mathematical terminology in the communication of ideas.

• *Everyday language and mathematical vocabulary are used to communicate geometric ideas.*

In newer curricular materials, children are encouraged to describe and compare shapes using vocabulary familiar to them. They talk about objects being "pointy," "tipped," or "going the wrong way." A particular rhombus might be described as a "squished square," and a cone called a "megaphone." They use words such as "corner" or "diamond," and the term "point" usually means vertex to them. Gradually, and as class discussions highlight particular features of shape, technical vocabulary is introduced.

Teachers exploring shape should also learn appropriate mathematical terms, e.g., "vertex," "side," "face," "edge," "surface area." Their work with vocabulary clarifies the purpose of precise mathematical terminology. For example, when children use the word "corner" applied to a rectangle or rectangular solid, they might be referring to any of several features. Using the words "vertex," "edge," or "angle" can eliminate ambiguity.

The word "same" is often heard in elementary mathematics discussions, but can have many meanings: Two rectangles might be said to be the same precisely because they are both rectangles, or because they are the same height, or because they are the same shape (even if of different size), or because one can exactly cover the other. On the other hand, "similar" is a common English word that has been given a precise mathematical meaning. (After a particular lesson on similarity, one teacher wrote, "I was surprised that 'similar' actually has a specific definition, that 'same' is less specific." Another confessed, "I thought I knew the definition, but when you [the instructor] put up rectangles that are similar, that blew my mind" [Bastable et al., in press].)

The role of mathematical definition takes on particular significance when children confront the contrast between definitions of triangle, rectangle, and square and their intuitions about which shapes are included in those categories. For example, a scalene obtuse triangle oriented so that no side is horizontal suddenly becomes a triangle, while a triangle-like object with rounded corners is now excluded.

In the following scene, a primary-grade teacher works with her students on squares and rectangles.

> **Scene 14, from a combined first/second grade classroom:** The teacher is leading a whole-group discussion about the children's definitions of rectangle. As the children talk, the teacher writes down their key points:
>
> • 4 sides
>
> • 4 angles
>
> • 4 corners

When the teacher asks, "So what's a square?" Roberto responds, "Four sides, four corners, four angles, and it's a square." The teacher writes:

- 4 sides

- 4 angles

- 4 corners

Clarisa: Actually, you don't need to say four corners and four angles; they're the same thing.

Teacher: Look at what we said about squares and what we said about rectangles.

Josh: Hey, they're the exact same thing! I'm thinking of shapes with the same definition but they're not a square or a rectangle.

He goes to the board to show what he's thinking of:

(based on Bastable et al., in press)

Here, the teacher helps her first and second graders see that the definition of rectangle is not simply a list of salient features, but must distinguish rectangles from all other shapes.

- *The process of measurement involves conceptual issues, not only procedures.*

Although most teachers understand the use of the ruler, few are aware of the conceptual issues involved in measurement. For example, they may not have considered explicitly the ambiguity of such words as "big" or "size," or that objects have several measurable attributes, e.g., that a box might be measured by its height, girth, surface area, volume, weight . . . Objects can be compared with respect to an attribute, either qualitatively (longer vs. shorter, heavier vs. lighter) or by assigning a numerical value, i.e., its measure. To measure an attribute, one must select an appropriate unit—for example, length must be measured by linear units, area by units that fill the plane, etc.—compare the unit to the object, and report the number of units. Also, perhaps implicitly understood, though often a surprise when explicitly remarked upon, smaller units yield larger counts.

Teachers should understand that measurements in the real world are approximations and how the unit chosen affects the precision of a measurement. They should be able to convert from one unit to another and use conversions to estimate measures. (For example, if the length of one's stride is approximately $2^1/_2$ feet and one paces a room to estimate its measure, one needs to multiply the number of paces by $2^1/_2$ in order to find the approximate measurement of the room in feet.) And they should know English and metric units of measure.

With regard to length, area, and volume, teachers must know what is meant by one, two, and three dimensions. A common misunderstanding is that perimeter is two dimensional since, it is argued, the perimeter of a rectangle has both length and width. Also, when asked to build a rectangular solid whose edges are double those of another, many teachers will double the length and the width, but not notice that a third dimension must also be doubled.

The following scene raises some conceptual issues involved in measuring area.

> **Scene 15, from a combined third/fourth grade classroom:**
> The teacher has asked her class to consider a triangle drawn inside a 3×4 rectangle with the following constraints: one side of the triangle is a side of the rectangle of length 4; the opposite vertex of the triangle is in the middle of the opposite side of the rectangle. What is the area of the triangle? The children copy the figure onto graph paper and, because of inaccuracies in their drawings, obtain different answers. Some children count up whole and partial units to 5; others, to 6. Emil, who believes the answer is 5, considers another way to view the problem.
>
> > Emil: If you add the squares outside the triangle and add it to the triangle, you should get 12. So, since $5 + 7 = 12$, you then want to get 7 on the outside.
>
> Omar picks up on the idea of looking outside the triangle and argues it should be 6.
>
> > Omar: The rectangle is worth 12. If you took the part that's not in the triangle and folded it up, it would make another equal triangle. The part in the middle is the same as the outside, so each part is 6.

   (based on Schifter et al., in press)

In Scene 15, third and fourth graders use what they know about partitioning shapes to figure out the area of a particular triangle. Future lessons will include experiences first with triangles inscribed in a 3×4 rectangle, then triangles inscribed in rectangles, encouraging the children to apply their own reasoning as they raise the level of generality. Rather than applying area formulas by rote, these children build a foundation for understanding why the formulas work.

However, both children and teacher in this classroom had already done much work on area prior to this lesson. Their ideas about area did not start here. In fact, understanding the area of a *rectangle* cannot be taken for granted among either children or adults.

Even to children who grasp the idea of using a plane-filling unit (e.g., a square) to measure area, the structure of a rectangular array is not obvious. They must coordinate unit squares, rows composed of unit squares, columns composed of those same unit squares, and a whole, the rectangle, which can be viewed as composed of squares, of rows, or of columns.

Although teachers may not face this conceptual challenge, there is still much for them to learn. They may remember the formula $A = L \times W$, but have no sense of how the linear units of the rectangle are related to the units that measure area or why multiplying the linear dimensions yields the count of those units (Simon & Blume, 1994). Similar issues apply to the volume of a rectangular solid (Heaton, 1992).

Once teachers have developed facility with composing and decomposing shapes, and possess an understanding of square units and rectangular arrays, they can analyze the behavior of measures under uniform dilations, first with respect to rectangles and rectangular solids, then generalizing to other shapes.

As Scene 15 suggests, an understanding of units of area, measuring rectangles, and composition and decomposition of shape prepare one to work through the logic of the area formulas for triangles, parallelograms, and trapezoids. An understanding of units of volume and the array structure of rectangular solids allows one to extend these ideas to the volume of other prisms and cylinders. Teachers should also know the formulas for the area and circumference of a circle, for the volume of cones and pyramids, and be familiar with some demonstrations of why they work.

Another subtopic worth considering is the complex relationships among different aspects of size. For many people, the impulse is to say that if A is bigger than B with regard to one attribute, the rest of A's attributes are bigger, too (Ball, 1991; Ma, 1999). However, one can have three rectangles, A, B, and C, where A is tallest, B has the largest area, and C, the largest perimeter; or three rectangular solids D, E, and F, where D is tallest, E has the largest surface area, and F, the most volume. Exploring the different aspects of size can be expanded into a major area of study (Fitzgerald & Shroyer, 1986), by posing such questions as: Which rectangles (with sides of whole-number lengths) can be made with a perimeter of 12 cm? of 20 cm? What is the area of each? Which shape maximizes area? Which rectangles can be made with an area of 24 square cm? etc.

Data Analysis, Statistics, and Probability

Statistics is the science of data, and the daily display of data by the media notwithstanding, most elementary teachers have little or no experience in this vitally important field. The statistics they need to know is naturally organized around the following three-step paradigm:

(1) Data production: designing studies to collect data relevant to questions of interest.

(2) Data analysis: using graphical representations, tables, and numerical summaries to find and describe patterns in the data.

(3) Interpretation: relating the results of data analysis back to original questions and stating conclusions; if necessary, designing and implementing a further study.

Probability serves as the theoretical basis for statistical inference, a subject beyond the scope of elementary school mathematics. But probability extends beyond its role in statistics and it is important to introduce teachers to the basic concepts enumerated below.

Summary of data analysis, statistics, and probability content.

- Designing data investigations:

  - understanding the kinds of questions that can be addressed by data.
  - creating data sets.
  - moving back and forth between the question (the purpose of the study) and the design of the study.

- Describing data:

    - describing shape: symmetry versus skewed data distribution and what this indicates about the question being addressed.
    - describing spread: range, outliers, clusters, gaps and what these indicate about the question to be addressed by the data.
    - describing center: mean, median, and mode and what these indicate about the question to be addressed by the data.
    - becoming familiar with different forms of data representation, e.g., line plots, stem-and-leaf plots, among others; recognizing that different forms of representation communicate different features of the data.
    - comparing two sets of data (not always of the same size).

- Drawing conclusions:

    - choosing among representations and summary statistics to communicate conclusions.
    - understanding variability.
    - understanding some of the difficulties that arise in sampling and inference.

- Developing notions of probability:

    - making judgments under uncertainty.
    - assigning numbers as a measure of likelihood.
    - becoming familiar with the idea of randomness.

- *Data studies can be designed to address particular kinds of questions.*

    Teachers begin by considering the question, What can be discovered through collecting, representing, and analyzing data?

    Data can be obtained for the elementary classroom through a variety of means. Data sets are available in the principal's office, from the town clerk, or through the Internet, for example. Data are collected in science experiments, lessons in measurement, daily lunch counts or other surveys.

    Yet any data collection procedure can present difficulties that must be addressed. What does one do with an incomplete data set? How does one account for inaccuracies in measurement? How can survey questions be framed to reduce ambiguity and produce analyzable data? What does one do with ambiguous data or data that do not fit predetermined categories? This very issue arises for the children in Scene 16.

> **Scene 16, from a third grade classroom:** In a unit called Project Huff and Puff, the children measured how far they could blow various objects and collected their results. One chart showed how far each child blew a styrofoam cylinder, but when the class looked it over, they saw Robbie listed twice. He had, indeed, done two trials, blown the cylinder 186 cm on his first try and 152 cm on his second. The children decided Robbie should be listed only once, and were discussing which number to use.

Erin: If we use the number in the middle, both are inside the middle. There is some of each number in the middle.

Andrea: You can count backwards from 186 to 152. (She writes down the numbers in descending order.) Now I can count how many numbers I just put down and then go halfway. I count down 17 from 186 and get 169.

Sydney: People don't want Robbie to get a high or a low score, and the average seems more fair.

Robbie: But I didn't get 169 as one of my distances. It wouldn't be true. It would be a lie!

(based on Russell et al., in press)

Rather than insist that Robbie should have followed directions and taken only one measurement, the teacher recognizes an opportunity for the class to deal with a common issue in data interpretation, how to work with data that don't fit the predetermined format. In considering this issue, the children come up against the idea of the mean, further discussed below.

Often, once data have been collected, one may find that the method of collection has not provided data that address the original question, or that it has, but another question emerges as more significant or interesting. In their own data investigations, teachers must consider how to use pilot data to revise their questions and data collection procedures.

• *The field of statistics offers a variety of methods to explore and describe data.*

Studying the shape of a data set and how data are spread has largely been neglected in school mathematics, but is critical to interpreting data (Shaughnessy, 1999). What does the shape of a data set indicate? Are the data clustered, spread out, symmetric, or skewed? Do most of the data cluster around a single value? Are they bimodal? flat? Where are there gaps? What is the range? Are there outliers? What does any of these characteristics indicate about the phenomenon represented?

Another core question arising from any data set—What is typical in this set?—introduces particular statistical measures: mean, median, and mode. What does each tell about the data? What else is important to know about the data in order to interpret their meaning?

These questions are considered by an elementary school class in the following scene.

Scene 17, from a fourth grade classroom: The children in the class measured their own heights and then collected heights from a first grade classroom. They met in groups to plan how to represent these two sets of data and created their representations. With a set of representations posted in the front of the room, the children discussed what they saw in the data:

Ilya: First graders are a lot smaller. The biggest first grader is 54 inches.

Suzanne:	A lot of first graders are near the heights of the smallest fourth graders. Not many have gotten up to the average height of a fourth grader in our class. Most fourth graders are 57 or 58 inches.
Robert:	The fourth graders have a larger range of heights than the first graders. . . . The fourth graders have a range of 14 numbers and the first graders have a range of 8 numbers.
Mike:	There are more first graders than fourth graders. There are 22 first graders and 21 fourth graders.
Teacher:	Does that tell you something you could say generally about all first graders and all fourth graders?
Mike:	No.
Teacher:	What about the other observations, that first graders aren't even as tall as an average fourth grader? Do you think that's typical or just these two classes?

The discussion continues with children's observations.

Teacher:	*(toward the end of class)* Can you give me a number that says how much taller a fourth grader is than a first grader, thinking about all the different ways you've thought about first grader heights and fourth grader heights?
Diane:	Ten inches, because the tallest fourth grader is 64 inches and the tallest first grader is 54 inches.
Trudy:	I think a first grader is about five inches smaller than a fourth grader. I found the median of fourth graders and first graders and I just subtracted.
Suzanne:	Five or four inches because the average first grader, the most common height, is 53 inches and the average fourth grader is 58 inches or 57 inches.
Lincoln:	I think to find the range between the first and fourth graders we need data from all the fourth grade classes and all the first grade classes.

(based on Russell et al., in press)

Having collected and represented their data, these fourth graders tell their teacher what they see. But if she is to understand their comments about range, overlap, clusters, median, and mode, she must know more about these concepts than formulas convey—for example, what does the median communicate that is not given by the middle of the range? (This particular question is important since many children confuse the two.)

Several of the children in Scene 16—Erin, Andrea, and Sydney—demonstrate some appreciation of what the mean is and how one can find the mean of two numbers. However, Robbie's confusion is widely shared, even among teachers: How can a value represent an entire set of data when that value does not appear in the data even once? To develop a deeper understanding of what the mean represents, teachers might be given such problems as: Create a data set of eight elements whose mean is 7. Create a second data set of eight elements, three above the mean and five below the mean.

After exploring shape, spread, and typicality, it is important to consider them in relation to one another. What does a single number represent and why is it useful? Within particular contexts, what else is important to know about a data set? Scene 17 illustrates children making reference to these characteristics to compare two data sets.

Data representation is a vibrant field offering a wide range of possibilities for data display. Although recognition of such possibilities is useful, fluency with those commonly used is imperative. Most teachers are already familiar with tables, bar graphs, and pie charts. But they should also be able to use line plots, stem-and-leaf plots, and box plots, among other methods of data display. By exploring the same data set in a variety of representations, teachers can consider how the latter communicate different aspects of the data and how one decides which form of representation to use for a given purpose. Specifically with regard to categorical data, how do different choices of category yield different views of the data?

One common source of confusion, for both children and teachers, is the difference between representations of value and representations of frequency (Bright & Friel, 1998; Konold & Higgins, in press). For example, given data on family size, one must distinguish between the number of members of each family versus the number of families of given sizes.

Outcomes that describe categories (gender, color preference, etc.) can be summarized by counts and plotted in bar graphs or pie charts. It is important to note that most of these summary statistics and descriptions of spread do not apply to categorical data.

• *Data provide a selected view into phenomena from which conclusions can be drawn.*

Once data have been collected, one must consider which conclusions can be drawn and how one communicates findings. Different sets of categories, modes of representation, and summary statistics give different views of the data and, thus, suggest different interpretations.

Issues of sampling and inference can be explored at this level in an informal way. Questions to be addressed include: Can a particular data set represent a larger population? What can data from one population tell you about another? For example, could the data about a single class generalize to all students of that age? Why or why not? What factors would need to be considered? What does it mean for a sample to be representative? How might a sample be biased? Why does the size of the sample matter?

These are questions that begin to arise in Scene 17. The teacher asks, "Does that tell you something you could say generally about all first graders and all fourth graders?" and again later, "Can you give me a number that says how much taller a fourth grader is than a first grader?" Lincoln (who earlier in the lesson expressed his sense that these particular first graders are exceptionally tall) suggests that, in order to make a general claim, they need to collect data from the other first and fourth grade classes. A teacher with a good understanding of sampling will be able to take advantage of this opening.

A related idea is that of variability. For example, in Scene 16, Robbie recorded two values for two attempts at "huff and puff." The teachers might ask, "If each of you were to try again, as Robbie did, would you get the same or a different number?"

Modern statistics is often thought of as a study of processes and the causes of variability. Students might study the time it takes to get through the lunch line, or classroom absences over a month, then consider causes of variability in the results they obtain. They might even suggest improvements to reduce variability.

Although establishing a relationship between two variables is not likely to be part of the elementary school curriculum, teachers should understand the idea of correlation and realize that correlation does not imply causality. Children with larger shoe sizes read better than those with smaller shoe sizes (there is a positive correlation between reading ability and shoe size among elementary school children), but the relationship is certainly not causal. For other variables, such as amount of sleep and academic performance, there may be causal links, but these can not be determined by simply tallying the sleep habits and grades of a few students.

Finally, data investigations are, by their very nature, open-ended. It may be possible to reach conclusions, but questions always remain. Furthermore, there are generally several correct routes to these conclusions. At the same time, it is important to dispel the common, naïve impression that "you can try anything you like and eventually something will work."

● *The concepts of probability provide ways of considering likelihood under conditions of uncertainty.*

> **Scene 18, from a second grade classroom:** Groups of children are thinking about outcomes from rolling dice. As they engage in their experiment of rolling two dice and recording outcomes, they are surprised to notice that "7 kept winning." Before moving on to their experiment of rolling one die, the class discusses their predictions and most children expect that one number will predominate, just as the 7 had when using two dice. Then, after having collected data on their one-die rolls, the class meets to look at the data together and discuss their observations.

```
                  x
                  x
                  x
                  x  x                         x
            x  x  x  x                 x  x  x  x  x  x
            x  x  x  x  x  x           x  x  x  x  x  x
      x  x  x  x  x  x  x  x  x        x  x  x  x  x  x
      x  x  x  x  x  x  x  x  x    x   x  x  x  x  x  x
      x  x  x  x  x  x  x  x  x  x  x  x  x  x  x  x  x
   1  2  3  4  5  6  7  8  9 10 11 12  1  2  3  4  5  6  7  8  9 10 11 12
```

Alyssa: The numbers on the one-die chart came out almost the same with no number winning. On the two-dice chart, the 7 beat the other numbers.

Teacher: When we made our predictions, many people thought 12 would win. Why do you think 12 has only two checks?

Allison: You can only get 12 when you get two 6s and that's hard to get.

Shelley: There's only one way to get a 12 and there is only one way to get a 2.

> As the class works on the number of ways to roll each number, Jonah observes, "It's like a football field," thinking about the yard lines. "The number of ways is the highest in the middle and it's smaller at each end." (based on Clark, 1996)

Children begin learning about probability by considering the likelihood of particular events in their daily lives (e.g., will it snow tomorrow?): impossible, certain, less likely, more likely. In Scene 18, children's experiments with dice introduce them to more precise concepts of probability: When rolling a single die, each outcome, 1 through 6, is equally likely. When rolling two dice, a 7 is more likely than a 12 because there are more ways to get a 7. In later years, the children will learn to assign numerical values to describe the likelihood of these events.

In order to teach them, teachers, too, must work on these ideas. Experiments with a balanced spinner can help them recognize that the probability of the spinner landing on a particular color is proportional to the area of that color. Experiments with coin tosses and dice can lead to an understanding that, given equally likely outcomes, the probability of a particular event is equal to the ratio of the number of outcomes defined by the event to the number of total possible outcomes.

In their work on probability, teachers must think through what randomness means. In 60 rolls of a die, each number will not necessarily appear exactly ten times. When flipping a coin five times, the result of HTTHH is no more likely than TTTTT.

The coin problem is based on the idea that one is starting with a fair coin which is tossed in a consistent manner. However, the type of problem with which one is usually confronted in applied work involves comparing observation to expectations. Thus, the problem could be turned around: Given that in five tosses we got the result TTTTT, should we look at the coin to see if it is fair before we make another bet?

A particularly challenging idea involves the difference between predicting an individual event and predicting a pattern of events. Teachers could work with simulation to explore this issue: When flipping a fair coin, how can one demonstrate that after a long run of heads, a tail is not "due"?

It must be emphasized that teachers should explore probabilistic concepts through physical simulation. The concepts of probability are difficult for almost everyone, and activities such as flipping coins, rolling dice, spinning spinners are the best means to learning. Later, even as teachers advance to computer simulation, physical experience should continue to be consulted.

Course Structures

In Chapter 2, it was recommended that prospective elementary teachers take at least 9 semester-hours of mathematics. Institutions will decide for themselves, preferably as collaborations between mathematics and education departments, how to structure the three courses to address the content described in this chapter. Some may choose to offer courses on number and operations, on geometry, and on algebra, respectively, selecting activities related to data to support content in each of these areas. Other institutions may choose other themes to define their courses, perhaps integrating content from number, algebra, geometry, and data into each of the three courses. In either case, content and methods courses should be coordinated.

Although research findings on teacher learning are still sparse, research on children's content learning can and should inform course design. For example, what is known about middle grades students' understanding of fractions has brought to light issues teachers urgently need to work on. (See, for example, Behr & Hiebert, 1988 or Carpenter et al., 1993.)

As content courses are designed, it should be kept in mind that teachers will not learn all the mathematics they need to know in their undergraduate studies—even if, from their instructors' perspectives, the course content has been covered. But if their undergraduate studies cultivate an interest in and capacity for mathematical activity, teachers will be prepared to continue learning in the context of their everyday practice. Furthermore, having developed in their undergraduate training a curiosity about mathematical ideas and an appreciation of mathematical pursuits, many more practicing teachers will be interested in continuing their mathematics studies—an interest these institutions should be prepared to address.

As for approaches to teaching mathematics to teachers, readers should be aware that several projects have been using "records of practice"—videotapes of elementary school mathematics lessons, print cases that illustrate children's mathematical thinking, samples of student work, teachers' reflections—as media of instruction (Ball & Bass, 2000a; Ball & Cohen, 1999; Lampert & Ball, 1998; Barnett et al., 1994; Schifter et al., 1999; Stein et al., 2000). These materials situate the mathematics in contexts resembling the elementary classrooms in which the subject matter is to be employed. For example, many of the scenes presented in this chapter are suitable bases for mathematics lessons for teachers. Thus, in courses taught by Magdalene Lampert, teachers take on for themselves the question posed by students in Scene 5: Given the diagram below, illustrating how three pieces of cheese are distributed among 16 mice, how do we name the portion each mouse gets?

Or, teachers could be asked to consider the question posed in Scene 3: Which features of the methods presented by Henry and Michaela to solve $728 \div 34$ reveal parallel lines of reasoning? Or again, they might work through the solutions offered by the children depicted in Scenes 1 and 2 to reconcile the different strategies.

By approaching content in this way, teachers who, themselves, have never had opportunities to work on the mathematics these children are addressing, now gain access to ideas that lay a foundation for more advanced work. On the other hand, for those who have already mastered this material, the cases provide a means to "decompress" it (in the language of Ball & Bass, 2000a), "to deconstruct . . . [their] own mathematical knowledge into less polished and final form, where elemental components are accessible and visible" (p. 98).

Useful as these "records of practice" might be to teachers learning mathematics, opportunities to work on mathematical questions of their own—to become curious, to offer conjectures, to devise proofs, and to solve problems—are indispensable. They need to become familiar with the pleasure of figuring things out, as well as with the concomitant experiences of confusion and frustration, tolerating their discomfort long enough for things to fall into place. If teachers never learn what this experience feels like, they won't have the gumption to allow their students to go through it, either.

For the instructor, this kind of practice entails as much listening and questioning —and gumption—as telling or explaining. If teachers are to become mathematical thinkers, they must be given opportunities to think—to go down blind alleys as well up productive avenues. The attentive instructor will look for strengths in the ideas teachers offer, pose questions to help them analyze their errors, and point them toward mathematically fruitful terrain.

For some instructors, teacher journal writing is an essential component of practice (Lampert & Ball, 1998). There, teachers record their problem-solving process and write about the mathematics they are learning—their new insights as well as the ideas with which they are struggling. Such writing allows teachers to consolidate their understanding and, not incidentally, also provides a valuable assessment mechanism for the instructor.

An Invitation to Mathematicians

This chapter began by observing that many prospective teachers have been less than successful mathematics students. Readers about to become their instructors may feel uneasy about how far back they'll have to reach to connect with these teachers. In going over the content described in this chapter, instructors may be tempted to start with topics in algebra because these seem respectably college-level. However, this is to begin at precisely the wrong place. If teachers have not had opportunities to make meaning for whole-number operations, to ground an understanding of why the operations work the way they do, ideas of generalization or function will elude them.

Taking on this work involves a different kind of mathematical challenge. The ideas teachers need to put together will seem so obvious to the mathematician as to be invisible. For example, why should anyone worry about why you can't additively decompose the divisor in a division problem? You can't do it because it doesn't work! The question seems absurd. But to the teachers who pose it, the question is compelling. The challenge to the mathematician is to figure out what makes this a substantive question and then to find a context in which teachers can think about it productively.

In one class where this issue arose, the teachers suggested that because making up word problems for particular numerical expressions had proved helpful before, they would try it here: What if there were 12 candy bars to share among 6 children, 4 girls and 2 boys? Each child gets 2 candy bars, $12 \div 6$. But if you decompose the divisor, 12 candy bars to share among 4 girls and 12 between 2 boys. . . . Suddenly, there were gasps and "oh's." "It's a different situation!" "It doesn't make sense to add the number of candy bars each girl and each boy get." The concrete context gave meaning to the symbols, offering access to the ideas of division. As one teacher later wrote, "Seeing the division example as a word problem was [mind-]boggling. Suddenly the 'why won't it work' appeared so clear" (Schifter, 1993).

In taking responsibility for the mathematics education of elementary teachers, mathematicians are invited, in effect, to re-enter the world of the naïve mathematical thinker. The recognition that the "unsophisticated" questions teachers pose do raise fundamental issues should inspire instructors to find contexts in which these can be addressed fruitfully. This means, at least initially, approaching the mathematics from an experientially based direction, rather than an abstract/deductive one. Isn't this the way each of us starts our individual journey into the world of mathematics?

References

Ball, D. L. (1989). *Videotape of classroom lesson, September 19, 1989.* Ann Arbor, MI: University of Michigan.

Ball, D. L. (1991). Research on teaching mathematics: Making subject matter knowledge part of the equation. In J. Brophy (Ed.) *Advances in research on teaching: Teacher's subject matter knowledge and classroom instruction* (Vol. 2, pp. 1–48). Greenwich, CT: JAI Press.

Ball, D. L. (1993). With an eye on the mathematical horizon: Dilemmas of teaching elementary school mathematics. *Elementary School Journal, 93*(4), 373–397.

Ball, D. L. & Bass, H. (2000a). Interweaving content and pedagogy in teaching and learning to teach: Knowing and using mathematics. In J. Boaler (Ed.), *Multiple perspectives on the teaching and learning of mathematics* (pp. 83–103). Greenwich, CT: JAI/Ablex.

Ball, D. L. & Bass, H. (2000b). Making believe: The collective construction of public mathematical knowledge in the elementary classroom. In D. Phillips (Ed.), *Yearbook of the National Society for the Study of Education, constructivism in education* (pp. 193–224). Chicago: University of Chicago Press.

Ball, D. L. & Cohen, D. K. (1999). Developing practice, developing practitioners. In L. Darling-Hammond & G. Sykes (Eds.), *Teaching as the learning profession: Handbook of policy and practice.* San Francisco: Jossey-Bass.

Ball, D. L. & Wilson, S. W. (1996). Integrity in teaching: Recognizing the fusion of the moral and the intellectual. *American Educational Research Journal, 33*, 155–192.

Barnett, C., Goldenstein, D., Jackson, B. (1994). *Mathematics teaching cases: Fractions, decimals, ratios, & percents.* Portsmouth, NH: Heinemann.

Bastable, V. & Schifter, D. (in press). Classroom stories: Examples of elementary students engaged in early algebra. In J. Kaput (Ed.). *Employing children's natural powers to build algebraic reasoning in the content of elementary mathematics.* Hillsdale, NJ: Lawrence Erlbaum Associates.

Bastable, V., Schifter, D. & Russell, S.J. (in press). *Examining features of shape: Casebook.* Parsippany, NJ: Dale Seymour Publications.

Battista, M. T. (1999). Fifth graders' enumeration of cubes in 3D arrays: Conceptual progress in an inquiry-based classroom. *Journal for Research in Mathematics Education, 30*(4), 417–448.

Battista, M. T. & Clements, D. H. (1998). *Seeing solids and silhouettes.* Menlo Park, CA: Dale Seymour Publications.

Behr, M. & Hiebert, J. (1988). *Number concepts and operations in the middle grades.* Reston, VA: National Council of Teachers of Mathematics.

Behr, M., Wachsmuth, I., & Post, T. (1985). Construct a sum: A measure of children's understanding of fraction size. *Journal for Research in Mathematics Education, 16*, 120–131.

Bright, G. W. & Friel, S. N. (1998). Graphical representations: Helping students interpret data. In S. P. Lajoie (Ed.), *Reflections on statistics: Learning, teaching, and assessment in grades K–12*. Mahwah, NJ: Lawrence Erlbaum.

Carpenter, T. P., Corbitt, M. K., Kepner, H. S., Lindquist, M. M., & Reys, R. E. (1981). *Results from the Second Mathematics Assessment of the National Assessment of Educational Progress*. Reston, VA: National Council of Teachers of Mathematics.

Carpenter, T. P., Fennema, E., & Romberg, T. (1993). *Rational numbers: An integration of research*. Hillsdale, NJ: Erlbaum.

Carpenter, T. P. & Levi, L. (1999). *Developing conceptions of algebraic reasoning in the primary grades*. Presented at the Annual Meeting of the American Educational Research Association, Montreal, Canada.

Carraher, D., Brizuela, B., & Schliemann, A., (2000). Bringing out the algebraic character of arithmetic: Instantiating variables in addition and subtraction. In T. Nakahara & M. Koyama (Eds.), *Proceedings of the 24th Conference of the International Group of the Psychology of Mathematics Education* (Vol 2., pp. 145–151). Hiroshima, Japan: Hiroshima University.

Clark, E. (1996). Patterns: Prediction and probability. In D. Schifter (Ed.). *What's happening in math class? Envisioning new practices through teacher narratives* (Vol. 1, pp. 14–17). New York: Teachers College Press.

Fitzgerald, W. & Shroyer, J. (1986). *The mouse and the elephant*. Palo Alto, CA: Addison Wesley.

Graeber, A. O. & Tanenhaus, E. (1993). Multiplication and division: From whole numbers to rational numbers. In D. T. Owens (Ed.), *Research ideas for the classroom: Middle grades mathematics* (pp. 99–117). New York: Macmillan.

Heaton, R. (1992). Who is minding the mathematics content? A case study of a fifth grade teacher. *Elementary School Journal, 93*(2), 153–162.

Heaton, R. (2000). *Teaching mathematics to the new standards: Relearning the dance*. New York: Teachers College Press.

Hiebert, J., Carpenter, T., Fennema, E., Fuson, K., Wearne, D. Murray, H., Olivier, A., & Human, P. (1997). *Making sense: Teaching and learning mathematics with understanding*. Portsmouth, NH: Heinemann.

Kaput, J. (Ed.). (in press). *Employing children's natural powers to build algebraic reasoning in the content of elementary mathematics*. Hillsdale, NJ: Lawrence Erlbaum Associates.

Konold, C. & Higgins, T. L. (in press). Statistics and data analysis. In J. Kilpatrick, W. G. Martin, & D. Schifter (Eds.), *Research companion to the Principles and Standards of School Mathematics*. Reston, VA: National Council of Teachers of Mathematics.

Lampert, M. & Ball, D. L. (1998). *Teaching, multimedia, and mathematics: Investigations of real practice*. New York: Teachers College Press.

Ma, L. (1999). *Knowing and teaching elementary mathematics: Teachers' understanding of fundamental mathematics in China and the United States*. Hillsdale, NJ: Lawrence Erlbaum Associates.

National Council of Teachers of Mathematics. (2000). *Principles and standards for school mathematics.* Reston, VA: Author.

Russell, S. J., Schifter, D. & Bastable, V. (in press). *Working with data: Casebook and facilitator's guide.* Parsippany, NJ: Dale Seymour Publications.

Russell, S. J., Smith, D. A., Storeygard, J., & Murray, M. (1999). *Relearning to teach arithmetic* (videotapes and study guide). Parsippany, NJ: Dale Seymour Publications.

Schifter, D. (1993). Mathematics process as mathematics content: A course for teachers. *Journal of Mathematical Behavior, 12*(3), 271–283.

Schifter, D., Bastable, V., & Russell, S. J. (in press-a). *Measuring space in one, two, and three dimensions: Casebook.* Parsippany, NJ: Dale Seymour Publications.

Schifter, D., Bastable, V. Russell, S. J. (with Yaffee, L., Lester, J. B., & Cohen, S.) (1999a). *Making meaning for operations: Casebook.* Parsippany, NJ: Dale Seymour Publications.

Schifter, D., Bastable, V. Russell, S. J. (with Lester, J. B., Davenport, L. R., Yaffee, L. & Cohen, S.) (1999b). *Building a system of tens: Facilitator's guide.* Parsippany, NJ: Dale Seymour Publications.

Schifter, D., Bastable, V., Russell, S. J. (with Cohen, S., Lester, J. B., & Yaffee, L.) (1999c). *Building a system of tens: Casebook.* Parsippany, NJ: Dale Seymour Publications.

Schifter, D., & Fosnot, C. T. (1993). *Reconstructing mathematics education: Stories of teachers meeting the challenge of reform.* New York: Teachers College Press.

Shaughnessy, M. (1999). *Research and assessment issues in the teaching and learning of probability and statistics: What questions might we be asking?* Presented to the Research Presession of the 1999 Annual Meeting of the National Council of Teachers of Mathematics, San Francisco.

Simon, M. A., & Blume, G. W. (1994). Building and understanding multiplicative relationships. *Journal for Research in Mathematics Education, 25*(5), 472–494.

Stein, M. K., Smith, M. S., Henningsen, M. A., & Silver, E. A. (2000). *Implementing standards-based mathematics instruction: A casebook for professional development.* New York: Teachers College Press.

Thompson, P. W. (1993). Quantitative reasoning, complexity, and additive structures. *Educational Studies in Mathematics, 25*(3), 165–208.

Tierney, C., Ogonowski, M., Rubin, A., & Russell, S. J. (1998). *Different shapes, equal pieces.* Menlo Park, CA: Dale Seymour Publications.

Usiskin, Z. (1987). Resolving the continuing dilemmas in school geometry. In M. M. Lindquist (Ed.), *Learning and teaching geometry, K–12, 1987 Yearbook of the National Council of Teachers of Mathematics* (pp. 17–31). Reston, VA: National Council of Teachers of Mathematics.

Usiskin, Z. (1988). Conceptions of school algebra and uses of variables. In A. Coxford & A. Shulte (Eds.), *The ideas of algebra, 1988 Yearbook of the National Council of Teachers of Mathematics* (pp. 8–19). Reston, VA: National Council of Teachers of Mathematics.

van Hiele, P. (1986). *Structure and insight.* Orlando, FL: Academic Press.

Chapter 8

The Preparation of Middle Grades Teachers

The mathematics needed by prospective middle grades[1] teachers encompasses the mathematics needed by teachers in the lower grades, but extended in several important ways to reflect the more sophisticated mathematics curriculum of the middle grades. Work with rational numbers and operations builds on earlier work with whole numbers and number operation knowledge. Concepts of symmetry and similarity depend on knowledge of shapes acquired in earlier grades. Developing a deeper understanding of measurement and new types of measures uses previously learned counting skills and their applications to finding areas and volumes of simple shapes. Graphing and interpreting both discrete and continuous data in a variety of ways follows graphing simple sets of discrete data in earlier grades. Middle grades teachers need to have a thorough understanding of the mathematics of the middle grades so that they can instill in their students the belief that they can make sense of the mathematics they are learning and the confidence to seek it. "Good mathematics learners expect to be able to make sense of the rules they are taught, and they apply some energy and time to the task of making sense" (Resnick, 1986, p. 191). Sense-making should be a theme in all courses for prospective teachers.

This chapter expands on the discussion in Chapter 4, and provides foundations and further explanation for the recommendations found there.

Teaching for Mathematical Reasoning in the Middle Grades

As middle grades students mature in their ability to undertake more complex mathematical learning, they are simultaneously developing their ability to reason more deeply about mathematics. Middle grades teachers need to have opportunities to come to understand the types of reasoning middle grades students are able to undertake, and then be able to challenge their students in ways that will lead them to reason and make sense of mathematics. They need to provide their students with opportunities to explore, conjecture, provide counterexamples, and justify—preparation for formal deductive reasoning in later mathematics classes. Reasoning and seeking understanding must be recognized by teachers as critical aspects of middle grades mathematics. This will not occur unless university mathematics instructors model for prospective teachers ways in which these aspects of mathematical learning can be commonplace in the mathematics classroom, and

---

[1]The term "middle grades" is used in this chapter, as in Chapter 4, to refer to Grades 5–8 and "secondary" to refer to Grades 9–12.

consciously make reasoning and understanding salient features of learning for their students.

The commonsense notion that students in the middle grades can perform deeper, more complex forms of reasoning than younger students is supported by the work of developmental psychologists (e.g., Case, 1985). Prospective teachers are frequently unaware of the developing ability of middle grades students to reason in progressively more complex ways. Yet mathematics is learned best if reasoning and problem solving play integral roles in the learning process. Teachers at all levels, including mathematics instructors of prospective teachers, need to understand the role of choosing appropriate tasks to further develop reasoning and problem-solving skills (Hiebert et al., 1997; Knapp et al., 1995; Stigler & Hiebert, 1999).

A very important form of mathematical reasoning that students should develop in the middle grades is the ability to reason about proportions, often referred to as proportional reasoning. It is therefore incumbent on teachers to understand how proportional reasoning develops and how this development can be promoted. Proportional reasoning has been called the "capstone of children's elementary school arithmetic" and the "cornerstone of all that is to follow" (Lesh, Post, & Behr, 1988, p. 94). Research has shown that children develop proportional reasoning in increasingly more sophisticated stages and that their development depends on their instruction (Lamon, 1995).

Students who are unable to reason proportionally will often approach proportion problems using an additive strategy (e.g., asked to predict the height of a tree in an enlargement of a photo in which a man is 2 inches tall and a tree 5 inches tall to one in which the man is 5 inches tall, will use the difference in the heights in the first photo to predict that the tree will be 8 inches tall in the second photo) rather than recognizing that the ratio of man to tree must remain constant. Progression from using only additive strategies (in this case the difference of two quantities) to recognizing and using both additive and multiplicative strategies appropriately is a hallmark of middle grades students' mathematical development. There are many situations in which multiplicative reasoning can be appropriately used, such as situations involving ratios, proportions, linear relationships, both multiplication and division and knowing when each operation is appropriate, compound units such as problems involving person-hours, and counting situations where the Fundamental Theorem of Arithmetic applies. One way of approaching difficulties with reasoning multiplicatively, which in many cases involves proportions, is to present a variety of situations in which a ratio can be appropriately used as an index of measure, such as in the photo problem. Prospective teachers' past introduction to proportion may have been restricted to setting up the standard $a/b = c/d$ equation and solving for one unknown missing value (e.g., if 2 balls cost \$3.00, how much will 7 balls cost? would be found by solving $2/3 = 7/x$). If so, they will benefit from encountering problems and questions that lead them to think more deeply about what makes a situation proportional in nature. Given time and tasks aimed at exploring the inconsistencies that can arise when differences rather than ratios are found and compared, prospective teachers will recognize multiplicative situations and will begin to understand some of the difficulties inherent in teaching ratio and proportion. Proportional reasoning is inherent in many areas of middle grades mathematics. For example, regression and correlation in statistics and both theoretical and empirical probability have their basis in proportional reasoning.

Other forms of reasoning also play important roles in middle grades curricula and thus need to be part of the middle grades teacher's own mathematical understanding. Reasoning about quantities and quantitative relationships can form a bridge from arithmetic to algebra. For example, if solving word problems is approached by first identifying the quantities (e.g., distance over, distance back, speed over, speed back, time over, time back) in word problems without attaching numerical values, and these quantities are then analyzed in terms of their relationships (e.g., in each case, the distance is equal to the speed times the time; the distances are equal), then either numerical values or algebraic variables can be used to ultimately represent the problem. This approach allows students to see solving algebra word problems as simply an extension of the kinds of problems they solved earlier. Few prospective teachers have learned to make this connection in their own schooling, and they benefit from attention paid to reasoning about quantities and their relationships.

Spatial reasoning, that is, mentally visualizing and reasoning about geometric objects and their relationships, should also be a focus of middle grades mathematics. Without opportunities to develop this type of reasoning for themselves, it is unlikely that prospective teachers will be able to assist their future students in reasoning about shapes in space. For example, for teachers who have not tried to do so, identifying two- and three-dimensional shapes that have rotational symmetry is very difficult, but with practice it becomes easier. Instructors of prospective teachers will find that their students have a range of abilities to visualize objects in space, and some will need more practice than others. Many prospective teachers are surprised to learn that others' spatial reasoning ability is much better or much worse than their own, and that spatial reasoning ability is not always closely correlated with other mathematical abilities. This information will help them be more responsive to their future students' abilities and needs for assistance. Software programs are available that can assist teachers and students to reason spatially, if accompanied by appropriate instruction.

Prospective teachers need to develop statistical and probabilistic reasoning not only to prepare for their future teaching in middle grades, but also because this type of reasoning is increasingly important in dealing with daily life. Margins of error are usually stated in surveys reported in newspapers and on television; what does this tell about the results of these surveys? Some surveys use a volunteer sample. Can valid interpretations be made? How does the choice of scale affect interpretations of graphs that appears in newspapers and magazines? If a person is fired from a job because he tested positive for drug use, is that action justified? What is the probability of a false positive test result?

The Mathematical Content Needed by Prospective Teachers

Mathematics coursework designed for prospective middle grades teachers must allow them to revisit the mathematics they learned in the past. But now the mathematics should be approached in a manner that will strengthen their understanding to the extent that they will not only be able to teach it to others, but they will also know when their students have understood and what to do if students have not understood. The coursework described here also contains ideas that will be new to prospective teachers—here too they must also achieve a level of understanding necessary for teaching this content to middle grades students. This chapter provides an overview of the mathematics content needed by prospective middle grades teachers so that they can lead their students to make sense of mathematics, to be able to communicate effectively about mathematics, and be able to use mathematics appropriately.

Four areas are discussed in some detail in this chapter:

- Number and operation.

- Algebra and functions.

- Measurement and geometry.

- Data analysis, statistics, and probability.

Coursework in these areas is expected to require about 12 of the 21 semester-hours of mathematics recommended in Chapter 2. The following sections are not intended as syllabi for courses but rather to provide indications of the type and level of coursework needed by prospective middle grades teachers. In keeping with the spirit of developing in-depth understanding of the mathematics fundamental to the middle grades, only these four areas of mathematics are discussed in the following sections. The final section discusses some options for the design of courses that include the topics described here, and further mathematics courses appropriate for a middle grades preparation program of 21 semester-hours.

Number and Operations

Summary of number and operations content.

- Develop a deep understanding of rational numbers and operations on rational numbers:

  - recognize that fraction symbols are used to represent a variety of mathematical situations.
  - understand decimal notation as an extension of place value.
  - estimate calculations with fractions, decimals, percents.
  - compare relative sizes of rational numbers.
  - recognize, among proposed solutions to rational number problems, those that are unreasonable.
  - determine which operation or operations can appropriately be applied to a situation.
  - understand percent as a special case of ratio.

- Understand the structure of the rational number system and the real number system:

  - change repeating decimals to fractions and fractions to decimals.
  - establish the relationships among whole, integral, rational, irrational, and real numbers.
  - understand the number line as a representation of the real numbers.
  - understand and use field axioms.

- Understand the mathematics that underlies standard algorithms:

  - use place value to explain multiplication and division algorithms for whole numbers and rational numbers expressed as decimals.
  - understand the mathematics that underlies commonly used algorithms for fraction operations.
  - understand the different ways of interpreting a division remainder and when each is appropriate.
  - make sense of computation strategies devised by students and appreciate the number sense involved in their creation.

- Understand and explain fundamental ideas of number theory as they apply to middle grades mathematics:

  - use the Prime Factorization Theorem and relate it to algebra.
  - be able to make conjectures about odd and even numbers and about composite and prime numbers, and provide justifications that prove or disprove the conjectures.
  - be able to justify and use the Euclidean Algorithm.

- Make sense of large numbers:

  - relate large numbers to known quantities (e.g., think about the relative size of a million, a billion, and a trillion by asking how long ago was a million, a billion, or a trillion seconds).
  - express and calculate with large and small numbers using scientific notation.

Discussion.

"Like common sense, number sense produces good and useful results with the least amount of effort. It is not mindlessly mechanical, but flexible and synthetic in attitude. It evolves from concrete experience and takes shape in oral, written, and symbolic expression. Links to geometry, to chance, and to calculation should reinforce formal arithmetic experience to produce multiple mental images of quantitative phenomena."   (National Research Council, 1989, pp. 46–47)

Strengthening rational number knowledge and rational number sense are absolutely essential components of middle grades mathematics teachers' preparation. Some prospective teachers think that they know all there is to know about rational numbers because they can carry out algorithms for computations with rational numbers and feel that they could teach these procedures to others. But rational number sense means being able to think flexibly about rational numbers—to attach meaning to the symbols we use to represent rational numbers, to be able to move easily among these representations, to understand and compare the relative sizes of a rational numbers, to be able to estimate the results of calculations involving rational numbers, to undertake many such calculations mentally, and to recognize unreasonable solutions to such calculations. For example, if asked for the result of $7/8 \div 1/4$, rather than automatically inverting and multiplying a person with number sense might ask, "How many fourths are in $7/8$?" then proceed to count them off:

three fourths in $3/4$, and then a half of a $1/4$ to reach $7/8$, so the answer is $3\ 1/2$. Some prospective teachers are "fraction avoiders" and habitually change all fractions to decimals before calculating. Although this behavior might not be harmful to them as individuals, as prospective teachers they will be less likely to develop the depth of understanding of fractions needed to teach in the middle grades. Reasoning with fractions (beyond learning algorithms for operations on fractions) has rarely been a curriculum topic in the background of prospective middle grades teachers, so it is not surprising that prospective teachers' knowledge of fractions is often limited to standard algorithms. Asking prospective teachers to write problems that can be solved by a particular arithmetic operation, for example, $2/3 \times 3/5$, helps them become aware that they need to know more than how to find the product of the two fractions.

Many people's understanding of fractions is limited to a part-of-a-whole interpretation, an interpretation that sometimes constrains their ability to think about fractions in other ways. For example, when asked to interpret the symbol $3/4$, many adults draw a circle, cut it into four equal parts and shade three (Silver, 1981). It is not surprising then that many middle grades students do not understand that a fraction symbol represents a quantity of something, that they do not interpret $3/4$ to mean 3 divided by 4, nor do they know that $3/4$ represents a point on the number line (Kerslake, 1986). Most have never thought of multiplication by $3/4$ as "shrinking" something to $3/4$ of its original size. An instructor of prospective teachers should ask these adults to interpret the symbol $3/4$ and find out how many of them can come up with these other meanings for the symbol $3/4$. Such an exercise will not only give the instructor insight into the understanding of these students, but can also be used as a basis for a class discussion of the meaning of the fraction symbol.

Admittedly, the part-whole relationship—3 of 4 parts of a whole as one way to think of $3/4$—is perhaps the most common way to interpret fraction symbols, and it is a useful one for thinking about fraction size, for example, that $8/9$ is a little less than 1, or that $5/9$ is a little more than half. This interpretation can then be extended to think about placement of fractions on the number line. Although comparing fractions in terms of size might seem trivial, only 35% of the eighth grade students taking the Seventh National Assessment for Educational Progress (Wearne & Kouba, 2000) were able to choose the correct ordering from smallest to largest of fractions such as $6/7$, $2/5$, and $1/2$. And in a study of middle grades teachers in a large midwestern city, only half correctly ordered $5/8$, $3/10$, $3/5$, $1/4$, and $1/2$ (Post, Harel, Behr, & Lesh, 1991). Both problems are quite simple for people with good number sense who are likely to think about these fractions in terms of their distance from 1, $1/2$, or 0 rather than using the procedure of finding a common denominator for the five fractions. (Of course, not all sets of fractions can be so easily ordered. Often students need only find a common denominator for a pair or two of the fractions to order the entire set.) Prospective middle grades teachers who are facile in using fractions will be more likely to help their future students develop this facility.

Too often instructors of prospective teachers don't spend sufficient instructional time on rational numbers because they mistakenly believe it to be review. There is evidence (e.g., Ball, 1990) that many teachers have never received instruction on operations on fractions beyond learning the algorithms and using them to solve simple word problems. Prospective middle grades teachers' knowledge of the mathematics they will be expected to teach is often superficial; they need to develop new mathematical understandings that will allow them to feel confident to teach rational number concepts and skills in a manner that will lead their students to develop robust understanding and strong skills. Courses for middle grades teachers should devote time to teasing out meanings for multiplication and division of fractions in particular. Answers to a problem such as "Is $3/4 \times 6/5$ greater than or less than $3/4$? Greater than or less than $6/5$?" can quickly suggest to an instructor whether or not a prospective teacher can think about fractions flexibly without needing to resort to algorithms that, in a case such as this, would give no insight into the question being asked. Prospective teachers need to think carefully about the unit to which each fraction in an operation refers, that is, in $2/3 \div 3/4 = 8/9$, the $2/3$ and $3/4$ each refer to some unit; $2/3$ of 1 and $3/4$ of 1, and the question asked is: How many $3/4$ of 1 are in $2/3$ of 1? But the referent unit for $8/9$ is the $3/4$; there is $8/9$ of one $3/4$ of 1 in $2/3$ of 1. Prospective teachers can learn this way of thinking about fractions and operations on fractions by identifying the referent unit for each fraction in word problems, such as "Miranda napped for $3/4$ hour. Shannon napped $2/3$ as long. How long did Shannon nap?" (What is the referent unit for each fraction, and for the solution?) Or, given a rectangle divided into 5 equal parts with 3 of the parts shaded, they can be asked questions such as: "Can you see $3/5$? If so, what is the unit? Can you see $2/3$? What is the unit? Can you see $3/2$? What is the unit? Can you see $5/3$ of $3/5$? What is the unit? Can you see $2/3$ of $3/5$? What is the unit?" (Thompson, 1995). Prospective teachers who work with manipulatives such as Pattern Blocks (available through many distributors) may find that they gain a great deal of insight into operations on fractions (and also on how concrete materials can be useful in teaching and understanding mathematics). Measurement situations lend themselves to working with fractions and with decimals.

Teachers usually introduce numbers represented as decimals in terms of their fractional equivalents and too rarely focus on extending place value understanding from whole numbers to decimals. But the middle grades student's understanding of how to represent a rational number as a decimal would be more useful if it were built on strong place value understanding developed for whole numbers in the earlier grades. When teachers themselves have developed a deep understanding of what decimal notation means in terms of place value, they have found that the additional time spent developing this understanding in their own students is gained back in later lessons focusing on comparing and operating on decimal numbers (e.g., Sowder, Philipp, Armstrong, & Schappelle, 1998; Wearne & Hiebert, 1988). This finding has implications for preparing middle grades teachers. Teachers also need to be aware of common conceptual difficulties middle grades students have when working with rational numbers. For example, some middle grades students think that 2.34 is larger than 2.4 because 34 is greater than 4; often they misplace the decimal point when multiplying decimal numbers. These difficulties are indications of the fragile knowledge of decimal notation that many middle grades students develop and never go beyond if their teachers do not have a good grasp of the role of place value in representing rational numbers using decimal notation and do not recognize students' difficulties in relating fractions, decimals, and percents.

Prospective teachers need to think carefully about the meaning of arithmetic operations, that is, determining which operation or operations will lead to a valid solution to a problem. Some exploration is needed, and often drawing diagrams can help teachers come to better understand operations. For example, some prospective teachers fall into the trap of thinking "Multiplication makes bigger and division makes smaller" (Graeber, Tirosh, & Glover, 1989). After multiplying to solve "If a pound of cheese costs $3.35, how much does 1.5 pounds of cheese cost?" some will then solve the problem "If a pound of cheese costs $3.35, how much does 0.89 pounds cost?" by dividing, because they know the answer must be less than $3.35 and think that they must divide to obtain such an answer. Nor do many prospective teachers recognize the difference between problems such as "When $7/8$ of a yard of ribbon is used to make 3 bows, how long is the ribbon in each bow?" and "When $7/8$ of a yard of ribbon is available to make bows each of which requires $1/4$ of a yard of ribbon, how many bows can be made?" The first problem embodies the "sharing" (or measurement or quotitive) interpretation of division while the second problem builds on the "repeated subtraction" interpretation of division. Some arithmetic exercises, such as $6 \div 1/2$, are usually easier if the repeated subtraction interpretation is used; how many halves are in 6 (e.g., how many times can $1/2$ be subtracted from 6)? Prospective teachers need to appreciate the difference in these interpretations in order to help their students make sense of division problems (Sowder, Philipp, Armstrong, & Schappelle, 1998). Research (e.g., Greer, 1992) has shown that multiplication and division situations are more complex than originally realized by curriculum developers and teachers. Recognizing how operations model situations demands a deeper understanding than simply viewing operations as computations.

Moving from fraction notation to decimal notation is straightforward given a good understanding of place value, but the reverse is more difficult. To have a full understanding of the rational number system, teachers must be able to move easily between these two representations of rational numbers. In particular, they need to understand how repeating decimals can be represented as fractions. (The notion that $0.99999\ldots = 1$ is a particularly difficult idea for prospective teachers to come to accept, even after an explanation of why this is so.) With the understanding of repeating decimals as a way of representing rational numbers, the question of non-repeating decimals should arise and can lead to a definition for irrational numbers.

By the end of fifth grade, students should have a good grasp of the multiplication and division algorithms used with whole numbers and decimal numbers, but not all do. The standard algorithms most of us learned were developed to be efficient, however complete descriptions may be lengthy and difficult to memorize without understanding. These algorithms are a challenge to teach because most teachers don't understand them themselves. An enormous amount of time in elementary and middle grades classrooms is spent on learning and practicing these procedures, especially that of long division, and is undertaken at the expense of other learning. The prospective teacher of middle grades needs to understand the rationale for the standard algorithms and how to teach these algorithms in a manner that will be understood and remembered by students without allowing the learning of these procedures to dominate the curriculum.

Prospective teachers and their students need to be able to interpret the answer to a long division problem in terms of the situation that led to the division. They often have particular difficulty dealing with remainders. When dividing 2664 by 84, is the answer 31 remainder 60? or $31 \, 60/84$? (or, as some children say, 31 remainder $60/84$?). Should the remainder be ignored, or should it be rounded to the next whole number? Should the division continue, resulting in a decimal number as a quotient, or should we stop here and present the remainder as a fraction? If we continue, what does "adding a decimal point and zeros" to 2664 mean? What now is being subtracted? Middle grades teachers should be helped to develop an appreciation of the role of such questions in leading to a deeper understanding of what quotients and remainders mean. Without this understanding silly but common mistakes occur such as, "11.5 buses are needed to take 322 children on a field trip, if 28 children can ride in each bus."

Although learning to use rational numbers demands a good portion of the middle grades mathematics curriculum, some whole number topics are also taught in these grades. Fundamental ideas of number theory, including the Prime Factorization Theorem, should be appropriately learned or reviewed by prospective teachers who have not had the opportunity to understand how these ideas prepare students for the learning of algebra. The Euclidean Algorithm is useful in finding the greatest common factor of two numbers, and discovering why it does so is a worthwhile exercise. Additionally, many number theory topics can be avenues that lead to practice with conjecturing and proving simple theorems about numbers, thus disproving for teachers the idea that proof is only relevant in geometry. Prospective teachers are sometimes surprised to learn that number theory has real-world applications, such as the manner in which large prime numbers are used in codes for electronic transfer of large sums of money.

Teachers should be able to work intelligently with very large numbers and very small numbers. Large and small are, of course, relative terms, but here we speak of those numbers that do not hold much meaning without developing some reference points, or benchmarks, particularly for large numbers. For example, "50,000 people would about fill the local stadium," "my city has a population of about 500,000," and "a million is a thousand thousands." If the national debt were paid off by charging every citizen an equal share, what would my share be? Rescaling in this manner offers another avenue to using proportional reasoning. Work with very large or very small numbers can motivate the need for and appreciation of scientific notation, which provides an effective and efficient way to symbolize and organize numbers. The structure of scientific notation needs to be understood, as does its relationship with significant digits. Teachers should be familiar with the manner in which scientific notation is represented on calculators.

Prospective teachers usually know how to carry out operations on integers, but are sometimes at a loss to explain the reasoning behind the algorithms. In particular, they cannot explain why "the product of two negatives is a positive." Multiplication of negative numbers has been called the first mathematical idea taught that cannot be explained by reverting to common sense, using models from the physical world. Teachers ought to know that mathematicians of the past also found it difficult to attach any meaning to negative numbers (Howson, 1996; Sfard, 2000). As work with rational numbers is extended to include negative rational numbers, teachers should extend their understanding of the number line to include numbers less than zero. After rational numbers and irrational numbers have been introduced, the real numbers can be defined and evidence provided that the number line can be used to represent both irrational and rational numbers. The rational numbers and the real numbers can now be discussed in terms of the field axioms, many or all of which would have been informally introduced previously and which

are of course very important for middle grades teachers to understand and be able
to use.

Mental computation and computational estimation should be included in the
curriculum for middle grades teachers because these forms of computation call on
and at the same time help develop the number sense these teachers need. Mental
computation and estimation both call for flexibility in moving from one representation to another, on appropriate rounding (for estimation), and on knowledge of
properties of operations. These skills need to be developed for whole numbers and
for rational numbers. Discussing, for example, how to estimate the sum of 384, 7,
and 6091 leads to a better understanding of significant digits; viewing 25×48 as
$100/4 \times 48$ or as $5 \times 5 \times 4 \times 12$ then as 20×60 makes this problem easier to compute
mentally. There is evidence that prospective teachers avoid mental computation
and computational estimation, and that in fact a very common way they approach
these problems when requested to do so is to visualize the paper-and-pencil algorithm and attempt to carry out those steps mentally (Levine, 1982; Sowder, 1989).
Developing flexibility with numbers through mental computation and estimation
leads prospective teachers and middle grades students to feel a degree of confidence
with numbers that they may never have experienced in the past. They also develop
a better understanding of field axioms.

Algebra and Functions

Summary of algebra and functions content.

- Understand and experience the different roles algebra plays:

  - as a study of patterns.
  - as a symbolic language useful in many areas of life.
  - as a tool for problem solving.
  - as the study of functions, relations, and variation.
  - as generalized arithmetic.
  - as generalized quantitative reasoning.
  - as a way of modeling and understanding physical situations.

- Develop a deep understanding of variables and functions:

  - relate tabular, symbolic, and graphical representations of functions.
  - relate proportional reasoning to linear functions.
  - recognize change patterns associated with linear, quadratic, and ex-
    ponential functions and their inverses.
  - draw and use "qualitative graphs" to explore meaning of graphs of
    functions.
  - understand the role of graphing calculators in the learning of algebra.

- Demonstrate skills connected to deep understanding:

  - represent physical situations symbolically.
  - graph linear, quadratic, exponential functions and their inverses and
    understand physical situations calling for each.

- solve linear and quadratic equations and inequalities.
- exhibit fluency in working with symbols.

Discussion. Algebra is a natural extension of arithmetic. It provides a symbolic language useful in representing quantitative relationships, it is a powerful tool for analyzing these relationships, and it provides models for decision making. Some educators describe algebra in terms of its roles in generalizing arithmetic reasoning and generalizing quantitative reasoning. When algebra is thought of as generalized arithmetic, its role as a language that encodes arithmetic properties is predominant. When thought of as generalized quantitative reasoning, algebra focuses on quantities as measurable aspects of a situation without necessarily attaching numerical values to those aspects. Both roles are useful. The first provides important links to what children already know. The second builds algebraic reasoning because it draws on aspects of growth and change, and because it focuses on relationships for the purpose of inference rather than for computation, thus providing a link to physical reality.

Prospective teachers whose previous algebra courses focused only on solving equations without understanding reasons for solution procedures and their relationships with arithmetic properties will understandably have a very limited notion of what algebra is about, and will be unequipped to address the curricular breadth now encompassed in school algebra. To prepare their students for future study of algebra and to be prepared themselves to teach algebra, teachers need to expand their own understanding of algebra. Coursework that allows students to begin to use algebra to model physical situations will allow them to see the usefulness of algebra as they become more skilled at expressing situations with symbols.

The teaching of algebra has been intensely examined over the past decade. Research studies have shown that the concepts of variable and function are poorly understood even by students studying calculus (e.g., see chapters in the book edited by Harel and Dubinsky, 1992). Thus it should not come as a surprise that prospective middle grades teachers often need experiences that will help them better understand functions. Computer technology and graphing calculators create new opportunities for them to learn about functions and graphing. Many educators say that the study of algebra ought to begin much earlier and extend throughout the grades, that algebra ought to be a curriculum strand rather than a course. There is no need to choose—algebra can be taught as a curriculum strand throughout the grades and as a course to follow arithmetic if that is the policy of the school district. The point here is that if prospective teachers think about algebra in its many roles they will begin to extend their understanding of algebra beyond its role as a body of procedures, and see how it can fit into many areas of the curriculum throughout the middle grades years. In addition, they should be prepared to teach a full year course in algebra, because some school districts are experimenting with requiring algebra in the eighth grade. Thus, they need to acquire both concepts and competencies in their algebra coursework.

One way of approaching the preparation of teachers of algebra is to enter this domain through the study of the mathematics of change. For example, situations can be presented in which it is necessary to plot and graph velocity as a function of time, distance as a function of time, velocity as a function of distance, acceleration as a function of time, and so on (see, e.g., Swan, 1985). This approach is quite new to most prospective teachers and forces them to come to grips with rate of change in different manifestations. "Qualitative" graphs allow them to focus on the variables and relationships as, for example, in drawing various graphs to represent a biker on flat ground traveling at a steady rate, then biking up a hill and slowing down, stopping at the top for a while, and speeding up when going down the hill, and finally slowing down at the bottom to a steady rate again, slower than when she approached the hill. Or graphs of functions without numbers on each labeled axis are given to students who then describe a situation portrayed by the graph. This type of study will prepare teachers to think more deeply about variables, functions, and graphs and what they represent.

Today, school algebra is often approached through investigations and problems, and the focus is on using algebra to make sense of the world around us. Both discrete and continuous quantities receive attention. Variables and relationships are identified in situations and used to model and understand the situations. There is a strong focus on being able to move easily from one representation to another—graphs, tables, and symbolic representations. Linear relations are given particular attention and pave the way for a later appearance of linear, quadratic, and exponential equations, and inequalities and their solutions. The study of linear functions should build on previous knowledge of proportional relationships. Teachers need to understand important ideas related to functions such as fixed points and asymptotes. They should understand how to use technology, particularly graphing calculators, to understand these ideas and how to illustrate them graphically, numerically, and symbolically. It is not that they will be teaching all of these ideas to their students, but rather that they need to know where what they teach leads. Technology is affecting our conceptions of what algebra is important to learn, and we must prepare teachers to think of algebra in different ways and to be able to use effectively a range of curriculum materials in their teaching.

Geometry and Measurement

Summary of geometry and measurement content.

- Identify two- and three-dimensional shapes and know their properties:

  - make conjectures about shapes and offer justifications for conjectures.
  - understand similarity and congruence of shapes.
  - be familiar with currently available software that allows exploration of shapes.

- Develop spatial reasoning through physical and mental activities:

  - manipulate mentally physical representations of two- and three-dimensional shapes.
  - determine the rotational and line symmetry for two- and three-dimensional shapes.
  - be familiar with interactive geometry software that allows movement of two- and three-dimensional drawings.

- Connect geometry to other mathematical topics, e.g., to:

- algebra via work with rectangular coordinate systems and with transformations.
- proportional reasoning via the study of similarity.

- Connect geometry to nature and to art.

- Understand measurement processes:

  - quantification of attributes of objects or ideas.
  - role of choice of measurement instrument and its influence on accuracy.
  - selection of unit of measure.

- Understand and use measurement techniques and formulas:

  - relate measurements within each of the two common systems of measure, English and metric.
  - estimate using common units of measurement.
  - develop and use formulas for measuring area and volume.
  - decompose and recompose non-regular shapes to find area or volume.
  - understand roles of π in measurement.
  - understand and use the Pythagorean Theorem.

Discussion. In current curricula for middle grades, students are expected to learn not only names of two- and three-dimensional shapes, but also to investigate characteristics of these shapes by visualizing, classifying, defining, conjecturing, and justifying or giving counterexamples to conjectures. A careful study of the meaning of congruence and of how congruence can be established should be included. Scaling is one approach to the study of similarity, which of course calls on proportional reasoning. Instructors of prospective teachers of middle grades will find that their students know names for two-dimensional shapes, remember (often rotely and sometimes incorrectly) some formulas for determining measurements of geometric shapes, and that some of their students remember standard proofs of theorems about triangles and parallelograms. One way to motivate prospective teachers to review two-dimensional shapes is to introduce three-dimensional shapes and describe them in terms of their two-dimensional faces. The names of less common three-dimensional shapes and identifying these shapes are often new to prospective middle grades teachers. The study of shapes should focus on properties and relationships of the shapes, and prospective teachers should be provided with opportunities to use software programs that allow for exploration of shapes to an extent not possible if the shapes must all be constructed with hand-held instruments. Circular shapes and shapes with circular cross-sections such as plates, cups, oatmeal boxes, and garbage cans lead to explorations of the special relationship of the perimeter and diameter of a circle. These types of explorations can lead to a level of conjecturing, proving, and disproving that enhances geometric reasoning and can lead to a deeper understanding of the role of proof in geometry.

The inability to reason spatially is a particular weakness of many prospective teachers who have never experienced a disciplined way of thinking about movement in space. For example, given different arrangements of six adjoining square regions, can they determine which arrangements could be folded along adjoining edges to form a cube? Or, can they envision how to slice a cube to get a cross-section that is a square? A non-square rectangle? An equilateral triangle? A trapezoid? Reasoning with two- and three-dimensional shapes can lead quite naturally to a study of geometric transformations and of symmetry. This background will allow, if an instructor so wishes, an excursion into the realm of tessellations of the plane and of space. Software that displays movement of three-dimensional objects can be especially useful in developing visualization ability.

Making connections between geometry and other areas of mathematics is an important aspect of preparing teachers to teach mathematics. The rectangular coordinate system, so often used to represent equations in two variables, lends itself well to investigating motions in the plane. The study of transformations allows prospective teachers insight into the role functions can play in geometry. The study of dilations leads to a study of proportions. Geometry should also be studied as it occurs outside of mathematics, such as in nature and in art. For example, a study of the "golden ratio" can lead to interesting explorations in art. Transformations also play a major role in artwork of many cultures—for example, they appear in pottery patterns, tilings, and friezes.

The different ways in which we measure attributes of geometric shapes is only one aspect of the study of measurement. The fact that we have found ways to quantify and measure so many aspects of our lives of work and of play is too little appreciated. Think about ways have people found to measure such things as blood pressure, atmospheric pressure, gum disease, the hotness of peppers, and the health of a new born child. Although teachers do not need to know the specifics of each of these ways of measuring, they should reflect on the role of measurement in advancing our knowledge. How do new forms of measure come about? Lord Kelvin (1824–1907), inventor of the Kelvin Temperature Scale, once said, "When you can measure what you are speaking about, and express it in numbers, you know something about it; but when you cannot measure it, when you cannot express it in numbers, your knowledge is of a meager and unsatisfactory kind." Prospective teachers can benefit from developing their own units for measuring quantities, such as sixth-grade students' answers and explanations to an open-ended problem given on a district test, to gain an appreciation of difficulties involved in developing measures.

Teachers in this country must understand, connect, and be able to teach both the English and the metric system of measurement. They need to understand the important role of having standard units of measure for each type of measurement, and how those units are related to units of other measures particularly in the metric system, which can be tied back to what students know about place value. They must also understand that measurements of continuous quantities are approximate, and that the need for greater or less accuracy influences the choice of the instrument selected for measuring. Teachers should learn to develop personal benchmarks for estimating common units of measure—the space from the shoulder to the fingertips of the opposite arm outstretched is usually about a meter but of course this unit will vary from person to person. A raisin weighs about a gram, a liter is slightly more than a quart, and so on. Prospective teachers should have opportunities to investigate relationships between types of measurements to think about how to design instruction that will lead to understanding measures, for example, how does holding area of rectangles constant affect perimeters of the rectangles? They need to have answers for questions such as: Why do the angles of a triangle always add up to 180°? What does π mean?[2]

Developing formulas for measuring area and volume should be undertaken in such a way that teachers realize that they can later redevelop a formula if it is not remembered. Activities involving surface area and volume can lead to better understanding of attributes and of units of measure. Decomposing complex geometric shapes into shapes that can be more easily measured is also a skill that develops with practice. Scaling and the study of similar figures call upon proportional reasoning. Situations calling for indirect measurement should be explored. The Pythagorean Theorem and at least one of its proofs, perhaps one that uses notions of composing and decomposing geometric shapes, should be included within a unit on measurement.

Many geometry activities, for example, problems involving transformations of two-dimensional figures, can be quite nicely presented via the coordinate plane, and illustrate the clear connection between geometry and algebra. The distance formula, based on the Pythagorean Theorem, offers a way of measuring the lengths of line segments in the coordinate plane.

The study of geometry and of measurement described here will prepare a teacher to be comfortable and proficient in teaching middle grades geometry as it now appears in many middle grades curriculum materials. It will not prepare teachers to teach a geometry course now typically offered in the secondary school, but prospective middle grades teachers should be familiar with the role of axioms, theorems, and proofs in the geometry curriculum of the secondary school. A course for middle grades teachers should include at least a brief foray into high school mathematics, at least to the extent that clarifies for middle grades teachers the geometry that follows and builds on that of the middle grades. Alternately, a course could be designed that prepares both middle grades and secondary teachers for the geometry appropriate in grades 7–12 curricula.

Data Analysis, Statistics, and Probability

Summary of data analysis, statistics, and probability content.

- Design simple investigations and collect data (through random sampling or random assignment to treatments) to answer specific questions:

  - formulate questions that can be addressed through data collection and interpretation.
  - make decisions on what and how to measure.
  - understand what constitutes a random sample and how bias is reduced.
  - understand how surveys are undertaken and what can be learned from them.
  - understand how statistical experiments are designed and what can be learned from them.

---

[2]See, e.g, Ma, 1999, p. 116.

- Understand and use a variety of ways to display data:

  - interpret and use bar graphs and pie charts for categorical data.
  - interpret and use histograms, line graphs, stem-and-leaf plots, and box plots for continuous data.
  - interpret and use scatter plots, regression lines, and correlations for bivariate data.

- Explore and interpret data by observing patterns and departures from patterns in data displays, with particular emphasis on shape, center, and spread:

  - recognize shapes of data distributions (e.g., symmetric, skewed, bimodal).
  - develop and use measures of center and spread.
  - identify misuse of cause-and-effect interpretations of correlations.

- Anticipate patterns by studying, through theory and simulation, those produced by simple probability models:

  - develop theoretical probabilities using information about equally likely outcomes.
  - develop empirical probabilities through simulations; relate to theoretical probability.

- Draw conclusions with measures of uncertainty by applying basic concepts of probability:

  - calculate and understand probabilities of independent and dependent events.
  - understand conditional probability and some of its applications.
  - understand sampling distributions, how they arise, and what can be learned from them.
  - understand margin of error and confidence intervals.
  - understand expected values.

- Know something about current uses of statistics and probability in many fields.

Discussion. Of all the mathematical topics now appearing in middle grades curricula, teachers are least prepared to teach statistics and probability. Many prospective teachers have not encountered the fundamental ideas of modern statistics in their own K–12 mathematics courses, and in fact need convincing that they need to learn this mathematics to be prepared to teach in the middle grades. Even those who have had a statistics course have probably not seen material appropriate for inclusion in middle grades curricula. Statistics is very much a context-rich subject. Interpretation, within the context of the problem, is very important. Prospective teachers must develop both skills for calculation and those for interpretation.

Teachers themselves need to learn to be critical consumers of data and statistical claims.

Currently, middle grades curricula emphasize exploratory analyses of data sets, some collected by the students and others found from outside sources such as the Web. The obvious bar graphs and pie charts for categorical data, which provide another opportunity to use proportional reasoning, are familiar to most prospective teachers. Moving from discrete data to continuous data motivates the need for graphs such as histograms, line graphs, stem-and-leaf plots, and box plots. After prospective teachers develop a few initial graphs on their own, much time can be saved by using a software program developed for teaching conceptual understanding rather than for undertaking tedious calculations. The time saved is better spent on learning to interpret the graphs and related summary statistics, such as the median, mean, and mode as measures of center and the interquartile range and standard deviation as measures of spread. Prospective teachers should have practice with and develop understanding of the role of conjecturing using sample data, and they should understand that conjectures as to why certain patterns appear in data are part of the exploratory process. They should encourage their future students to think about data in this way.

Association between two variables, represented by two univariate sets of data, is basic to most statistical investigations. To show possible associations, two-way tables are used for comparing the conditional distributions of two categorical variables; parallel box plots for comparing the shapes, centers and spreads of two (or more) measurement variables; and scatterplots, with attendant regression lines and correlations, for studying the bivariate relationship between two measurement variables. Lines of best fit allow for predictions and also connect with properties of linear functions learned in algebra. Examples of the misuse of statistical association to make cause-and-effect statements, particularly in the case of regression and correlation, can be brought into class discussions.

A statistical investigation is designed to answer specific questions posed by the investigator, and sound answers depend upon a well-formed question and a well-designed study that pay careful attention to randomization and proper measurement procedures. Teachers should understand that sample surveys involve the random selection of samples from a fixed and well-defined population for the purpose of estimating parameters of that population, such as the proportion of students in the school who plan to vote in the coming student elections. In sampling, teachers and students must have a clear understanding of the difference between a sample (and its statistics) and a population (and its parameters); of what constitutes a random sample; of the role of random selection in reducing bias in a survey; of the role of random assignment in reducing confounding of variables in an experiment; of why stratification may be important in the sample design, of how to determine a stratified random sample; and of how sample size is related to the precision of results. Experiments involve the random assignment of treatments to experimental units for the purpose of drawing conclusions about possible treatment differences, such as the possibility that water stays hot longer in Styrofoam cups than in paper cups. In experimenting, teachers and students must have a clear understanding of what constitutes a treatment and what constitutes an experimental unit; of how one randomly assigns treatments to experimental units; of how to set up a matched pairs experiment; and of why the possibility of replication (repeating each treatment on more than one experimental unit) is important.

Probability can be thought of as a way of thinking about the future. It is a way of describing the chance that an outcome will occur. Predictions can be made based on probabilities found theoretically (as can be the case of the probabilities of the various outcomes from tossing a fair die) or empirically (as in the case of collecting data by actually tossing a balanced die). Theoretical probabilities can be checked by collecting data to see if the observed relative frequencies agree with their theoretical models. The notion of a model providing a theoretical probability that can be compared to empirical results is fundamental to the study of the relative frequency concept of probability that is most useful to the study of statistics. The fact that, under random sampling, the empirical probabilities actually converge to the theoretical (the law of large numbers) can be illustrated by technology (computer or graphing calculator) so that an understanding of probability as a long-run relative frequency is clearly established. For example, simulations of the probability of getting a 5 when two balanced dice are tossed allows prospective teachers to visualize what happens to the empirical probability of the event after, say, 10 tosses, 100 tosses, and 1000 tosses.

The rules for calculating probabilities of compound events made up of independent or dependent events need many examples and much discussion by prospective teachers before they are fully understood. Tree diagrams and area models are quite helpful to many. This understanding requires a discussion of conditional probability, a topic rich with applications that leads prospective teachers to realize the usefulness (and the subtlety) of probability. For example, suppose 5% of the general population uses drugs and a drug test used is 95% accurate; then a person with a positive test result has only a 50-50 chance of actually being a drug user. Problems such as "If you know a family has two children, and a boy answers the door, what is the probability that the other child is a boy?" lend themselves nicely to simulation using coins or random numbers. Setting up a simulation can lead to better understanding of probability and the power to predict events. One goal of the course should be to consider carefully the many serious misconceptions people have about chance events, such as the "gambler's fallacy" (if a coin comes up heads for seven tosses in a row, it is certain to come up tails on the next toss), or the representativeness misconception (a hand of four cards of all queens is less likely that a hand with the 8 of hearts, the 2 of clubs, the 10 of clubs, and the 3 of diamonds).

As important as the actual calculation of probabilities for single events is, an understanding of probability distributions and how they arise, both empirically and theoretically, will lead to a much clearer understanding of likely and unlikely outcomes. For example, under the assumption 50% of a population will vote in the next election, the outcome of 40 voters in a random sample of 50 is unlikely not because its probability is small but, rather, because it is far out in the tail of the probability distribution of possible outcomes. (All individual outcomes have small probabilities in this scenario.) Distributions of sample statistics, like the proportion of voters, are called sampling distributions and are the basic building blocks of statistical inference. Summary measures for probability distributions are couched in terms of expected values, hence prospective teachers need some experience with expected value problems. Teachers in particular should have some understanding of the normal distribution, because much of the information they receive on testing is based on this distribution.

For properly designed studies, conclusions generally take the form of an estimate of a parameter or a decision as to whether to reject an hypothesized model. These forms of statistical inference have measures of error attached, either through the margin of error in an estimate or the chance of rejecting a true model. How far a sample statistic (like a proportion or mean) is likely to be from the target parameter can be assessed by looking at many different simulated sampling distributions of the statistic, under differing assumptions about the parameter and, perhaps, differing sample sizes. Teachers should understand the process of making inferences through simulated sampling distributions (which can be done effectively in middle grades) and its relationship with more mathematically based inference procedures taught at higher levels.

Proportional reasoning plays a crucial role in understanding both statistical and probabilistic ideas. Probabilities represent ratios, and dealing with ratios calls upon proportional reasoning. For example, if a mechanic says that he gets far more cars of type x to repair than cars of type y, does that mean the cars of type x are inferior? How can this problem be thought of as a proportion problem?

A Program to Prepare Middle Grades Teachers

Recommendation 11 of Chapter 2 advocates that mathematics in middle grades, from Grade 5 on, be taught by mathematics specialists. In recent years many states have developed credential programs for middle grades teachers, but some universities have been slow to instantiate them. Elementary teachers seldom have a mathematics background appropriate or sufficient for teaching middle grades mathematics, and most secondary programs do not examine the middle grades mathematics content in a manner needed by teachers of these grades. For example, neither elementary nor secondary teachers are likely to have examined the conceptual underpinnings of the rational number system in the manner needed to teach this topic in middle grades. Prospective middle grades teachers need coursework designed with their needs in mind. This coursework might also serve other populations, depending on how the courses are structured.

The actual manner in which these courses are designed will depend on how courses are already (or in the process of being) designed for elementary and for secondary teachers. Courses for middle grades teachers should strengthen these prospective teachers' own knowledge of mathematics and broaden their understanding of the mathematical connections between one educational level and the next. The following is an example of one 12 semester-hour option in which courses are designed to meet the needs of multiple audiences.

> A course offered for elementary teachers that focuses on understanding numbers and place value would be a valuable course to require of middle grades teachers, who not only need to understand the fundamental role of place value in the decimal number system and computations with rational numbers expressed as decimals, fractions, and percents, but also need familiarity with the curriculum of grades K–4.

A full semester of study of geometric shapes and their properties together with measurement of shapes and other quantities might serve both elementary and middle grades teachers, or a one-semester course could be designed to serve both middle grades and secondary teachers.

A course that focuses on conceptual understanding of probability and statistics (rather than on statistical procedures) might already be available among course offerings, but if not, one should be designed for this population or for middle grades and secondary teachers.

A course that examines the structural properties of integer, rational, real, and complex number systems and the ways in which algebraic reasoning and number theory are used in contemporary applications of mathematics could serve both middle grades and secondary teachers.

These four courses would include all the basic content thus far discussed. Many examples of appropriate problems and tasks can be found on the Web site of the National Council of Teachers of Mathematics, in the Principles and Standards of School Mathematics. New examples are being added to this Web site on an ongoing basis.

Additional coursework that allows prospective middle grades teachers to extend their own understanding of mathematics, particularly of the mathematics they are preparing their students to encounter, will also be required. We suggest that this second type of coursework contain at least one semester of calculus if a course exists that focuses on concepts and applications. (The usual calculus designed for engineers and mathematics majors would probably not have this focus because of its emphasis on connections with physics and engineering.) Number theory and discrete mathematics can offer teachers an opportunity to explore in depth many of the topics they will teach. A history of mathematics course can provide middle grades teachers with an understanding of the background and historical development of many topics in the middle grades curriculum. A mathematical modeling course, depending on the level and substance of the course, can provide prospective teachers with understanding of the ways in which mathematics can be applied. If the prospective teachers are likely to teach algebra, then coursework in linear algebra and modern algebra would be appropriate. If, in addition, the teachers might be expected to teach a full-year course in geometry, then they should have the same geometry coursework as prospective secondary teachers. These last options might require more than 21 semester-hours.

It is recommended that mathematicians designing courses for middle grades teachers work closely with colleagues from the college or school of education to develop a program for middle grades specialists. These programs are likely to vary from institution to institution. Currently, in some cases, the education faculty offer a methods course for middle grades mathematics teachers, but in other cases prospective teachers are offered only an elementary methods course or a secondary methods course. In such cases, mathematicians might be able to fill in some of the gaps by courses designed for middle grades teachers. Ideally, these mathematics courses will be combined with opportunities, arranged with colleagues in education, to observe middle grades teaching and to tutor small groups of middle grades students.

References

Ball, D. L. (1990). Prospective elementary and secondary teachers' understanding of division. *Journal for Research in Mathematics Education, 21*, 132–144.

Case, R. (1985). *Intellectual development: Birth to adulthood.* Orlando, FL: Academic Press.

Graeber, A. A., Tirosh, D., & Glover, R. (1989). Preservice teachers' misconceptions in solving verbal problems in multiplication and division. *Journal for Research in Mathematics Education, 20*, 95–102.

Greer, B. (1992). Multiplication and division as models of situations. In D. A. Grouws (Ed.), *Handbook of research on mathematics teaching and learning* (pp. 276–295). New York: Macmillan.

Harel, G., & Dubinsky, E. (Eds.) (1992). *The concept of function: Aspects of epistemology and pedagogy* (MAA Notes no. 25). Washington, DC: MAA.

Hiebert, J., Carpenter, T. P., Fennema, E., Fuson, K. C., Wearne, D., Murray, H, Olivier, A., & Human, P. (1997). *Making sense: Teaching and learning mathematics with understanding.* Portsmouth, NH: Heinemann.

Howson, G. (1996, July). *Mathematics and common sense.* Paper presented at the International Congress for Mathematics Education, Seville, Spain.

Kerslake, D. (1986). *Fractions: Children's strategies and errors.* Windsor, England: NFER-NELSON.

Knapp, M. S. and associates. (1995). *Teaching for meaning in high-poverty classrooms.* New York: Teachers College Press.

Lamon, S. J. (1995). Ratio and proportion: Elementary didactical phenomenology. In J. T. Sowder & B. P. Schappelle (Eds.), *Providing a foundation for teaching mathematics in the middle grades* (pp. 167–198) Albany, NY: SUNY.

Lesh, R., Post, T., & Behr, M. (1988). Proportional reasoning. In J. Hiebert & M. Behr, (Eds.), *Number concepts and operations in the middle grades* (pp. 93–118). Hillsdale, NJ: Erlbaum, and Reston, VA: National Council of Teachers of Mathematics.

Levine, D. R. (1982). Strategy use and estimation ability of college students. *Journal for Research in Mathematics Education, 11*, 350–359.

Ma, L. (1999). *Knowing and teaching elementary mathematics: Teachers' understanding of fundamental mathematics in China and the United States.* Mahwah, NJ: Lawrence Erlbaum Associates.

National Research Council. (1989). *Everybody counts.* Washington, DC: National Academy Press.

Post, T. R., Harel, G., Behr, M. J., & Lesh, R. (1991). Intermediate teachers' knowledge of rational number concepts. In E. Fennema, T. P. Carpenter, & S. J. Lamon (Eds.), *Integrating research on teaching and learning mathematics* (pp. 177–198). Albany, NY: SUNY Press.

Resnick, L. B. (1986). The development of mathematical intuition. In M. Perlmutter (Ed.), *Perspectives on intellectual development: The Minnesota Symposia on Child Psychology* (Vol. 19, pp. 159–194). Hillsdale, NJ: Erlbaum.

Sfard, A. (2000). Limits of mathematical discourse. *Mathematical Thinking and Learning, 2*(3), 157-189.

Silver, E. A. (1981). Young adults' thinking about rational numbers. In T. R. Post & M. T. Roberts (Eds.), *Proceedings of the Third Annual Meeting of the North American Chapter of the International Group for Psychology in Mathematics Education* (pp. 149–159). Minneapolis, MN: University of Minnesota.

Sowder, J. T. (1989). Affective factors and computational estimation ability. In D. B. McLeod & V. M. Adams (Eds.), *Affect and mathematical problem solving: A new perspective* (pp. 177–191). New York: Springer Verlag.

Sowder, J. T., Philipp, R. A., Armstrong, B. E., & Schappelle, B. (1998). *Middle grades teachers' mathematical knowledge and its relationship to instruction.* Albany, NY: SUNY Press.

Stigler, J., & Hiebert, J. (1999). *The teaching gap.* New York: Free Press.

Swan, M. (1985). *The language of functions and graphs.* Nottingham: Shell Centre for Mathematics Education.

Thompson, P. W. (1995). Notation, convention, and quantity in elementary mathematics. In J. T. Sowder & B. P. Schappelle (Eds.), *Providing a foundation of teaching mathematics in the middle grades* (pp. 199–221). Albany, NY: SUNY Press.

Wearne, D., & Hiebert, J. (1988). A cognitive approach to meaningful mathematics instruction: Testing a local theory using decimal numbers. *Journal for Research in Mathematics Education, 19*, 371–384.

Wearne, D. & Kouba, V. L. (2000). Rational numbers. In E. A. Silver & P. A. Kenney (Eds.), *Results from the Seventh Mathematics Assessment of the National Assessment of Educational Progress* (pp. 163–191). Reston, VA: National Council of Teachers of Mathematics.

Chapter 9

The Preparation of High School Teachers

There is a seductive plausibility to the "trickle down" philosophy underlying most current programs for the mathematical preparation of high school mathematics teachers—the idea that coursework in a standard mathematics major develops the reasoning skills necessary to teach high school mathematics well. This intuitively reasonable proposition has not been supported by research on teacher effectiveness. For example, Begle (1979) found that students' mathematics performance was unrelated to either the number of college mathematics courses their teachers had taken or the teachers' average grade in such courses. The number of courses taken by teachers was actually negatively correlated with their students' performance in 15% of the studies. Begle also found that whether a teacher had majored in mathematics had a statistically significant impact in only 20% of cases studied. More recently, Monk (1994) found that each of the first four college mathematics courses taken by high school mathematics teachers was associated with a 1.4% increase in their students' scores on an achievement test. But further mathematics courses and majoring in mathematics had negligible impact on teachers' effectiveness. The number of mathematics education courses taken by teachers had a positive impact on their students' achievement—comparable to that of the first mathematics content courses.

When content knowledge is measured by course-taking, this modest correlation between mathematics teachers' content knowledge and students' learning has been confirmed in other disciplines. In a review of related studies Darling-Hammond, Wise, and Klein (1995, p. 24) found "positive relationships between education coursework and teacher performance [that were] stronger and more consistent than those between subject-matter knowledge and classroom performance." Researchers at the National Center for Research in Teacher Education (NCRTE) found that teachers who majored in the subject they were teaching often were not more able than other teachers to explain fundamental concepts of their discipline (NCRTE, 1991, p. i). Their investigations led to the conclusion that, "teachers need explicit disciplinary focus, but few positive results can be expected by merely requiring teachers to major in an academic subject. Studying subject matter in relation to subject matter pedagogy helps teachers be more effective. Teacher education programs that emphasize the underlying nature of the subject matter . . . more often result in knowledgeable, dynamic teachers with transformed dispositions and understandings of subject matter and pedagogy."

Other research has helped to describe such knowledge more explicitly. It appears that instead of (or perhaps in addition to) acquiring knowledge of advanced mathematics, what effective teachers need is mathematical knowledge that is organized for teaching—deep understanding of the subject they will teach; awareness of persistent conceptual barriers to learning; and knowledge of the historical, cultural, and scientific roots of mathematical ideas and techniques (Ma, 1999; Shulman, 1986).

The case for rethinking the content preparation of high school mathematics teachers is strengthened by considering changes in high school curricula. Responding to the growing role of data analysis, statistics, probability, and discrete mathematics in science, engineering, computing, and business, new high school curricula have broadened the typical high school curriculum to include generous amounts of material from statistics and discrete mathematics. Other proposals have suggested that new calculator and computer technologies—with powerful graphic and computational tools—will transform the traditional emphases of high school algebra and geometry courses (Heid, 1997).

Expectations for high school teaching are changing too. Research on teaching and learning suggests that carefully designed instruction that, for example, engages students in collaborative investigations rather than passive listening to their teachers, will produce deeper learning and better retention of mathematics as well as improved social and communication skills (Bransford, Brown, & Cocking, 1999; Springer, Stanne, & Donovan, 1999). Calculator and computer tools have suggested new ways of teaching school and collegiate mathematics, encouraging laboratory-style investigations of key concepts and principles.

All of these changes in high school curricula and teaching challenge conventional patterns of teacher education. To make intelligent curricular decisions for their students and to teach current school curricula, future high school teachers need to know more and somewhat different mathematics than mathematics departments have previously provided to teachers. Because they are being urged to teach in different ways, prospective teachers also need to experience learning mathematics in those ways themselves.

To meet these needs, the education of prospective high school mathematics teachers should develop:

- Deep understanding of the fundamental mathematical ideas in grades 9–12 curricula and strong technical skills for application of those ideas.

- Knowledge of the mathematical understandings and skills that students acquire in their elementary and middle school experiences, and how they affect learning in high school.

- Knowledge of the mathematics that students are likely to encounter when they leave high school for collegiate study, vocational training or employment.

- Mathematical maturity and attitudes that will enable and encourage continued growth of knowledge in the subject and its teaching.

Responding to the Challenge

Chapter 5 gave two main recommendations for ways in which mathematics departments can attain these goals. First, the content and teaching of core mathematics major courses can be redesigned to help future teachers make insightful connections between the advanced mathematics they are learning and the high school mathematics they will be teaching. Second, mathematics departments can support the design, development, and offering of a capstone course sequence for teachers in which conceptual difficulties, fundamental ideas, and techniques of high school mathematics are examined from an advanced standpoint.

As mathematics departments explore ways to respond to the needs of future high school teachers, they will naturally be concerned about the compatibility of teacher preparation initiatives and needs of other students majoring in mathematics. At the time this report is being written, the MAA Committee on the Undergraduate Program in Mathematics (CUPM) is engaged in a comprehensive review of the mathematics major curriculum. That study will give special attention to the needs of prospective teachers. Preliminary CUPM work is revealing that the mathematical needs of prospective teachers have more in common with those of other students majoring in mathematics than many faculty realize. Many mathematics courses that were originally designed with a focus on preparation for graduate school now serve a constituency of undergraduates who major in mathematics but do not plan to attend graduate school, as well as undergraduates from other majors. Thus, most of the suggestions here that are designed to better serve the needs of prospective teachers seem likely to appear in some form in the final CUPM report. The following outline of mathematics and supporting courses is one way to provide core knowledge for future high school teachers while satisfying many requirements in a standard mathematics major.

Year One: Calculus, Introduction to Statistics, supporting science.

Year Two: Calculus, Linear Algebra, and Introduction to Computer Science.

Year Three: Abstract Algebra, Geometry, Discrete Mathematics, and Statistics.

Year Four: Introduction to Real Analysis, Capstone, and mathematics education courses.

These courses can be enhanced in ways that will make them more useful for future teachers and other undergraduates as well. The capstone course idea is of particular value to future teachers. But its aim of giving broad historical and cultural perspectives, insight into mathematics learning, and applications of technology should serve other mathematics majors too. The balance of this chapter is an elaboration of the ideas outlined in Chapter 5—giving some specific examples of ways that standard mathematics major courses can be redesigned to be particularly useful for future teachers and themes that can be developed in the capstone sequence for teachers. The following sections describe goals for high school mathematics teacher preparation in five areas that correspond to major strands of the high school curriculum—algebra and number theory, geometry and trigonometry, functions and analysis, statistics and probability, and discrete mathematics. Each section indicates conceptual connections among those strands, and the broad mathematical reasoning processes and the historical perspectives, that are of importance to teachers.

Algebra and Number Theory

Current school mathematics curricula connect algebra to topics in functions and analysis, discrete mathematics, mathematical modeling, and geometry. Graphing calculators, spreadsheets, and computer algebra systems encourage and facilitate those connections, and raise deep questions about the appropriate role of skill in algebraic manipulation.

To be well prepared to teach current high school curricula, mathematics teachers need:

- Understanding of the properties of the natural, integer, rational, real, and complex number systems.

- Understanding of the ways that basic ideas of number theory and algebraic structures underlie rules for operations on expressions, equations, and inequalities.

- Understanding and skill in using algebra to model and reason about quantitative relationships in real-world situations.

- Ability to use algebraic reasoning effectively for problem solving and proof in number theory, geometry, discrete mathematics, and statistics.

- Understanding of ways to use graphing calculators, computer algebra systems, and spreadsheets as tools to explore algebraic ideas and algebraic representations of information, and in solving problems.

Prospective teachers will enter undergraduate studies with some technical skill in algebra. They can manipulate familiar types of polynomial, rational, and exponential expressions to solve equations and inequalities and to transform given expressions into equivalent forms. They also have rudimentary knowledge of number theory concepts like primes, factors, multiples, greatest common divisors, and least common multiples, and they are proficient in operations with fractions, decimals, radicals, and complex numbers.

However, prospective teachers tend to have only limited understanding of the algebraic properties that characterize the various subsets of the complex number system, that justify methods for solving equations and simplifying expressions, and that guide operations in many other systems like polynomials, sets, logic, functions, or matrices. Their understanding of and skill in working with linear systems, matrices, and polynomials is often superficial.

Algebra and number theory in courses for mathematics majors. Typical programs for preparation of high school mathematics teachers have several components that contribute to their knowledge of algebra and number theory.

Calculus. Calculus courses are an excellent place to polish and extend all undergraduates' algebraic understanding and skill and to strengthen their facility with model building. Solving calculus problems requires a great deal of algebraic manipulation, and understanding related theorems can require further generalization and abstraction (for example, being able to write a particular power series and to determine whether or not it converges requires some algebraic manipulation. Being able to express a general power series requires abstraction and generalization of that process). Calculus also develops a more abstract sense of algebra and functions by expressing methods and results, such as integration by parts and the chain rule, in general notation. The algebraic aspects of calculus create many opportunities where an instructor's timely observations can give undergraduates new insights into algebra.

Linear algebra. After prospective teachers' skill and understanding with algebra has been enhanced in the relatively concrete setting of calculus, the stage is set to develop some theoretical understanding in a sophomore linear algebra course. The standard course in linear algebra gives extensive experience with linear systems and matrices, connections of those algebraic ideas to the geometry of vectors and transformations, and the beginnings of understanding and skill in algebraic proof. Most linear algebra courses strike a balance between the familiar $\mathbf{R}^2$, $\mathbf{R}^3$ and associated matrices and the more general $\mathbf{R}^n$, abstract vector spaces, and linear transformations. For prospective teachers, it is particularly important for vector space ideas to be well understood in $\mathbf{R}^2$ and $\mathbf{R}^3$, which are central to high school mathematics, before moving to abstract vector spaces. The challenge is that linear algebra courses have an increasing number of topics to cover. The study of $\mathbf{R}^2$ and $\mathbf{R}^3$ as vector spaces is a natural setting in which to develop students' understanding of connections between linear algebra and analytic geometry. One way to develop this area further is to devote several classes to applications of matrix-based analytic geometry in the growing area of computer graphics and visualization. This subject involves a variety of linear transformations of $\mathbf{R}^3$ onto $\mathbf{R}^3$ and projections of $\mathbf{R}^3$ onto $\mathbf{R}^2$. For example, an extensive amount of three-dimensional analytic geometry, including cross products, is needed to determine which sides of a given tetrahedron are visible to a viewer at a certain position in space. Another important application of linear algebra that is especially useful for teachers is the role of projections in pseudoinverses and least-squares curve-fitting. Regression lines, which are based on least-squares projections, are becoming part of the statistics taught in high school. However, they are also of fundamental importance to undergraduates in many fields that require linear algebra.

Abstract algebra. Most mathematics majors take only one semester of abstract algebra, so it is important that such a course develop students' appreciation of the breadth and power of algebraic structures. The last comprehensive CUPM report on the mathematics major (MAA, 1981/1989) suggested a syllabus for abstract algebra that offers one way to balance depth with breadth. It proposed what might be called "visiting the algebraic structure zoo": Spend the first two or three weeks of the course in a tour of the whole zoo to get a basic introduction to all the major species. Then spend most of the rest of course studying a few algebraic species in some depth. At the end of the discussion of each algebraic structure devote time to connections with other areas of mathematics and to applications in other disciplines.

For all undergraduates, but especially for future high school teachers, such an abstract algebra course can effectively build on familiar algebraic structures encountered in high school and other college mathematics courses. Examples of rings, integral domains, and fields are familiar from high school, and the most useful for future high school teachers. The amount of time spent on group theory would be less than is often the case in the first semester of a two-semester algebra course and group theory would be closely connected to concrete examples such as isometry groups. There are now some abstract algebra texts for such a syllabus. A second semester of algebra, geared for, among others, mathematics majors considering graduate school, can focus on groups and deeper topics such as Galois theory.

It is natural for such study of algebraic structures to include exercises that explore the algebraic properties that underlie numerical and symbolic operations in school mathematics. It is important for prospective teachers to understand how most extensions of the number system, from natural numbers through complex numbers, are accompanied by new algebraic properties, and why the field axioms are so critical for arithmetic. Algebra courses also should make prospective teachers aware of the many algebraic structures that they have encountered but may not have noticed, such as the isometry groups of regular polygons.

Number theory. Number theory has always been a popular elective in the mathematics major, especially among prospective teachers. Numbers are the most familiar of mathematical objects. The subject is a concrete setting for strengthening algebraic and proof-building skills. It is useful here again to explicitly examine mathematics underlying number theory concepts used in school mathematics. For instance, students can be asked to use unique factorization and the Euclidean Algorithm to explain familiar procedures for finding common multiples and common divisors of integers. Modular arithmetic provides mathematically rich examples of algebraic structures that can be explored by students at various levels of sophistication from elementary and middle school through high school. There are also a number of contemporary applications of number theory to problems in coding and computing that high school students find accessible and intriguing.

Algebra and number theory in the capstone sequence. There are at least five themes that could be addressed in a capstone look at algebra and number theory.

Historical perspectives. Prospective teachers need to develop an eye for the ideas of mathematics that will be particularly challenging for their students. One very useful guide to such topics is the historical record showing how the ideas were first developed (Boyer, 1991; Katz, 1992, 2000; Swetz et al., 1995). The history of algebra is rich in stories that illuminate the challenge of expressing number patterns and properties in modern notation, of developing algorithms for solution of equations, of making sense of fractions, negative numbers, irrational numbers, and complex numbers.

For example, there is evidence that Babylonian mathematicians over 4,000 years ago were able to solve quadratic equations equivalent to $x^2 - x = 870$ using a method that was a special case of the quadratic formula now in common school use. However, the problems and solutions were expressed in rhetorical form and base 60 numeration. It was not until the 17th century that such problems and solutions were expressed in something like the modern symbolic notation that secondary school students are now expected to master quickly. Furthermore, it was not until modern times that mathematicians were comfortable with the general quadratic $x^2 + px + q = 0$ where p and q are positive, because such equations can have negative or complex roots.

Investigation of these historical issues involves substantial mathematics. It also provides prospective high school mathematics teachers with insight for teaching that they are unlikely to acquire in courses for mathematics majors headed to graduate school or technical work.

Common conceptions. Although historical analysis shows the difficulties encountered in development of fundamental algebraic ideas and techniques, there are many aspects of algebra that seem to be persistent challenges in learning the subject. Experienced teachers come to know where students are likely to make mistakes in algebraic manipulation and reasoning. It makes sense to address these issues in the preparation of high school teachers and to analyze the mathematical and psychological factors that lead students to common errors. For example, it is not at all uncommon for students to overgeneralize algebraic laws like the distributive property to conclude that $(a + b)^2 = a^2 + b^2$, to be puzzled by expressions like 0^0, or to have only limited understanding of the rationale underlying methods for solving equations and systems of equations. There is a rich body of research on the range and roots of such algebraic difficulties (Confrey, 1990; Kieran, 1992). Collaboration of mathematicians and mathematics educators could produce capstone activities that would prepare high school teachers with deep understanding of the issues from both mathematical and psychological perspectives.

Applications. Because algebra occurs throughout all branches of mathematics, it is easy for students and prospective teachers to get the impression that it is a tool in the service of other topics. Many applications of algebra do occur in the context of problems in geometry or analysis, but there are some very important applications that apply core algebraic topics directly. For example, linear programming problems make an excellent setting for illustrating the usefulness of core algebra topics like linear equations and inequalities. Cryptography is a direct application of both abstract algebra and number theory. Aspects of both linear programming and cryptography can be developed at levels that are appropriate for and interesting to high school algebra students.

Technology. College algebra and number theory courses are beginning to exploit and study the use of calculator and computer tools for doing and learning algebraic ideas. Calculators with powerful computer algebra systems (CAS) are now available at prices that make them accessible, and they are used more and more in schools. With only modest instruction, a high school student can use such tools to do most of the calculation problems that fill course assignments and examinations in algebra. The implications of this new-found calculating power are still being worked out in a variety of formal and informal classroom and curriculum experiments. High school teachers whose careers will cover the next 30 or 40 years need experience with these tools in their own learning and problem solving that will usefully inform their teaching. There is growing evidence that use of CAS technology can expand high school algebra and that intelligent users need common sense and mathematical habits of mind to use these tools wisely.

In addition to CAS software, computers offer other tools that require algebraic understanding. For example, spreadsheets are one of the most widely used computer tools, and designing a useful spreadsheet requires flexible ability to express numerical relationships in algebraic notation. College mathematics courses not often incorporate spreadsheet tasks. However, high school computer literacy courses often introduce students to this tool, and there are important ways that spreadsheet work can illuminate important mathematical ideas. For example, spreadsheet formulas often implicitly define recursive procedures, so spreadsheets can be used to explore fundamental properties of arithmetic, geometric, and more general sequences and series.

The capstone sequence is a perfect place to explore these emerging technologies and their role in high school algebra curriculum and teaching. In addition to providing a new perspective on familiar algebraic concepts and techniques, the capstone can consider important issues about when and how technology should be used to maximum benefit. Once again, collaboration of mathematics and mathematics education faculty in designing this course would help to make examination of this perspective productive for future teachers.

Connections. Because algebra is the language in which so many relationships are expressed throughout mathematics, it is a natural topic to use in exploring important mathematical connections. Although it is tempting to assume that undergraduates will see those connections throughout their mathematical studies, experience suggests that a reflective overview of the mathematical landscape is of value.

For example, a first course in linear algebra only begins to develop an understanding of the interplay among matrices, systems of equations, and vectors or of applications like least-squares curve-fitting in data analysis, Markov processes in probability, or dominance matrices in graph theory. The first course in abstract algebra often uses symmetry groups of polygons as examples of groups, without connecting more generally to the geometry of transformations. Most abstract algebra courses do not connect solution of polynomial equations in radicals to the impossibility of classical Greek construction problems like doubling the cube. Each of these topics is rich in mathematics and in opportunities to give prospective teachers deeper understanding of the scope and important processes of the subject they will teach.

Geometry and Trigonometry

High school geometry was once a one-year course of synthetic Euclidean plane geometry that emphasized logic and formal proof. Recently, many high school texts and teachers have adopted a mixture of formal and informal approaches to geometric content, de-emphasizing axiomatic developments of the subject and increasing attention to visualization and problem solving. Many schools use computer software to help students make geometric experiments—investigations of geometric objects that give rise to conjectures that can be addressed by formal proof. Some curricula approach Euclidean geometry by focusing primarily on transformations, coordinates, or vectors, and new applications of geometry to robotics and computer graphics illustrate how mathematics is used in the workplace in ways that are accessible and interesting to high school students.

To be well-prepared to teach the geometry in high school curricula, mathematics teachers need:

- Mastery of core concepts and principles of Euclidean geometry in the plane and space.

- Understanding of the nature of axiomatic reasoning and the role that it has played in the development of mathematics, and facility with proof.

- Understanding and facility with a variety of methods and associated concepts and representations, including transformations, coordinates, and vectors.

- Understanding of trigonometry from a geometric perspective and skill in using trigonometry to solve problems.

- Knowledge of some significant geometry topics and applications such as tiling, fractals, computer graphics, robotics, and visualization.

- Ability to use dynamic drawing tools to conduct geometric investigations emphasizing visualization, pattern recognition, conjecturing, and proof.

Most prospective mathematics teachers enter undergraduate study familiar with basic properties of two- and three-dimensional figures (mainly triangles, quadrilaterals, circles, and related solids), and they have some facility in constructing proofs of elementary results for planar figures (generally relationships involving congruence, similarity, parallelism, and perpendicularity). Through pre-calculus study they will have gained some skill in elementary coordinate geometry and vector techniques. Despite the fact that measurement of geometric objects is a core subject in precollege mathematics, it is quite likely that even strong entering undergraduates will have only a formula-driven understanding of that topic.

Geometry and trigonometry in undergraduate mathematics courses.
The standard calculus and linear algebra courses for mathematics majors give students extensive experience with important geometric ideas and techniques— especially coordinate methods, vectors, transformations, and trigonometry. But the focal point of geometry for prospective high school teachers is usually a course in college geometry. There are several ways that these courses can help to prepare high school mathematics teachers.

Calculus. The analytic geometry component of current calculus courses is substantially smaller than a generation ago, with much of the familiar content moved into an already crowded pre-calculus course where it is not treated as fully. The traditional calculus topics of analytic geometry and trigonometry that have always been important for teachers and engineers are now of growing importance for computer scientists as well. It seems appropriate to restore some of that important material to a more central position in the college curriculum. Such a move would be very helpful in the preparation of high school teachers.

Linear algebra. One of the many ways in which linear algebra can be viewed is as generalized analytic geometry. There is an inherent tension between geometry and algebra in a linear algebra course, as well as between $\mathbf{R}^n$ and abstract vector spaces. The key is to maintain a sensible balance. It is also important to constantly view questions from both points of view, and to translate algebraic results into geometric language and vice versa. Attention to this interaction of geometry and algebra will be very useful for future high school teachers.

Geometry. Upper-division geometry courses for teachers typically involve re-examination of Euclidean geometry from a axiomatic point of view along with rudimentary non-Euclidean geometry. Some versions of this course include or even start from transformational or coordinate points of view, but the most typical approach is synthetic reasoning in the spirit of Euclid, with modern standards of rigor.

A major goal of a collegiate geometry course should be to deepen prospective teachers' understanding of standard Euclidean theorems and principles and their skill in use of axiom-based reasoning. A careful review of high school geometry can have substantial and valuable conceptual content. However, prospective teachers should also be acquainted with other aspects of geometry, in order that they understand that geometry is not restricted to axiomatic geometry and are prepared to teach high school topics such as computer graphics. Such topics also provide an opportunity to strengthen undergraduates' geometric and algebraic skills.

A geometry course for teachers could also examine the geometry of the sphere (Henderson, 2000) and the shapes of useful figures like conic sections. Another topic that often interests prospective teachers is the development of the geometry of congruence and similarity from axioms about isometries and similitudes. This development can be connected to the algebra of matrices and complex numbers in ways that appeal to high school students as well as prospective teachers.

Computer graphics provides a new viewpoint for re-examining a variety of topics in synthetic and analytic geometry. Artistic notions about perspective must be translated into mathematics in order for such perspectives to be represented on a computer screen. Dynamic geometry software permits experiments with geometric constructions that provide opportunities for students and teachers to explore the visual world mathematically. The graph theory component of discrete mathematics also contains several interesting topics of a geometric nature, such as planar graphs. Graph theory also provides an interesting approach to the classical topic of Platonic solids via Euler's formula.

A high school teacher who has some familiarity with aspects of modern geometry such as tiling and fractals and with applications such as computer graphics and robotics will convey a richer view of the subject to students. Fitting all of those topics into one college geometry course that also treats gives an in-depth axiomatic development of Euclidean geometry runs a clear risk of covering ground without developing depth of understanding. As is the case with abstract algebra, it seems promising to survey some topics quickly and then treat a selected few in depth.

Geometry and trigonometry in the capstone sequence. Knowledge of geometry for teaching can be provided in the proposed capstone sequence. This kind of course can explicitly trace the historical development of key ideas, identifying questions that were challenging for mathematicians and will be difficult for students. It can show ways that development of deep ideas can be started in high school courses, and it can examine thoughtfully the interplay of intuitive, exploratory work and axiomatic proof. The capstone sequence may also be an ideal setting for re-examination of key trigonometric ideas to assure that prospective teachers have the depth of understanding that is essential to effective curricular decision making and instruction. Some specific examples illustrate the mathematical issues that can be addressed in such a course.

Historical perspective. College geometry courses often develop key ideas with some attention to the chronology of events beginning in Greek mathematics and culminating in formulation of non-Euclidean geometry during the 19th century. However, there is really much more to the subject than time allows in a typical formal geometry course. Tracing the roots of the Pythagorean Theorem alone reaches back to developments in Babylonian, Egyptian, and Asian mathematics and highlights the dramatic influence of Greek focus on axiomatic methods. Exploration of the geometric approach to number and algebra that dominated Greek mathematics will provide future high school teachers with geometric representations of algebraic ideas that can be used to help their students.

The visual side of geometry makes it an excellent place to explore the interplay of mathematics and cultural traditions. The visual arts of nearly every ancient and contemporary culture embody important geometric concepts and principles.

Common conceptions. Geometry doesn't have the collection of procedural skills associated with arithmetic or algebra, but there are still some important and predictable conceptual difficulties in learning the subject. Students have trouble with the logical issues involved in proof—often failing to understand the power of universally quantified statements, the role of counterexamples, the connections among propositions and their converses and contrapositives, and the class inclusions associated with definitions for parallelograms, rectangles, and squares. The connection between visual exploration of ideas and formal definitions or proofs is a persistent source of learning difficulties, with students often unable to free their minds from perceptions such as the belief that a pair of perpendicular lines must always consistent of one horizontal and one vertical line. The natural human tendency to think in terms of linear proportionality very often leads students to difficulty in understanding the effects of similarity transformations on perimeter, area, and volume.

Teachers who understand these common mathematical and psychological challenges in learning geometry are prepared to plan instruction that heads off or corrects such student misunderstandings. Collaboration by mathematicians and mathematics educators in the design of a capstone course could help prospective teachers lay the foundation for development of this kind of deep content knowledge for teaching.

Applications. The connection of plane and solid geometry to objects in the physical world makes applications of fundamental principles very easy. However, the standard college geometry course seldom focuses on those topics that are so useful in the resource kit of high school teachers. The capstone sequence could profitably address both classical and modern applications of basic geometry—from the ubiquitous Pythagorean Theorem, the conic sections, and simple mechanics to modern topics in computer graphics, fractals, and robotics.

Technology. In the same way that computer algebra systems can be used to explore symbol manipulation, computer drawing tools provide powerful aids in geometric explorations. Tasks that require construction of figures with given properties help teachers (and will help their students) see the logical interdependence of key ideas. The dynamic grab-and-drag features of computer drawing tools illustrate the universality of theorems in a way that goes far beyond typical paper and pencil explorations.

For example, after constructing a triangle and its medians with software like *Cabri Geometry* or *Geometer's Sketchpad*, one can grab a vertex and drag it across the screen to form an infinite number of new triangles. If the medians are constructed so that they satisfy the definition of median (rather than simply being drawn in approximately the correct position), all three will continue to intersect at the same point. Moreover, the ratios of the lengths of the two segments in which each median is cut will remain 2:1, even though the lengths of the medians themselves will vary. These kinds of geometric explorations lay a foundation for high school students' understanding of formal proof. They also illustrate an important aspect of creative mathematical work and of the way in which software can embody a mathematical definition. The capstone course can help make future teachers comfortable with use of such tools and also address the connections between experimental and deductive mathematics.

Connections. In the same way that algebraic notation provides a kind of universal language for representing quantitative information, geometric shapes provide visual representations of ideas. The connection between geometry and algebra provided by coordinate methods is among the most powerful tools of mathematics. The connection works both ways: Algebra allows computational methods to be used in reasoning about geometric objects, but geometry provides helpful visual images for algebraic calculations, limiting processes in calculus, reasoning about probability, and display of data in statistics.

Although there is a geometric basis for the subject of trigonometry, right-triangle, and periodic-function aspects of that topic have been traditionally taught in a separate high school course and as part of pre-calculus studies. This may be one reason why prospective high school mathematics teachers often have some technical proficiency in the trigonometry of right triangles when they come to undergraduate studies, but lack deep understanding of the core geometric principles that make trigonometry possible. The capstone sequence is a natural place to re-examine trigonometric ideas, several key identities (law of sines, law of cosines, Pythagorean Theorem), the addition formulas, and the general notion of identity—and to make or reinforce connections with geometry.

Functions and Analysis

The concept of function is one of the central ideas of pure and applied mathematics. For nearly a century, recommendations about school curricula have urged reorganization of high school mathematics so that study of functions is a central theme. Computers and graphing calculators now make it easy to produce tables and graphs for functions, to construct formulas for functions that model patterns in data, and to perform algebraic operations on functions. Prospective high school mathematics teachers must acquire deep understanding of the function concept in general and the most important classes of functions: polynomial, exponential and logarithmic, rational, and periodic. For functions of one and two variables, teachers should be able to:

- Recognize data patterns modeled well by each important class of functions.

- Identify function types associated with various relationships like $f(xy) = f(x) + f(y)$, or $f'(x) = kf(x)$, or $f(x + k) = f(x)$.

- Identify the types of symbolic representations associated with each class of functions and the way that parameters in those rules determine particular cases.

- Translate information from one function representation (tables, graphs, or rules) to another.

- Use function representations and operations to solve problems in calculus, linear algebra, statistics, and discrete mathematics.

- Use calculator and computer technology effectively to study individual functions and classes of related functions.

Students who study pre-calculus mathematics in high school or college will probably encounter the function concept and the major function families in a serious way. This introduction needs conscious reinforcement and extension in college mathematics courses.

Functions in courses. Future high school teachers meet functions in calculus, linear algebra, and various other courses of the mathematics major. It is easy for those encounters to focus on the procedural aspects of function notation, operations, and graphs. High school and college students often come to see functions as nothing more than alternate ways of expressing algebraic information. There are several ways that standard courses can convey a broader view of functions and their role throughout mathematics.

Calculus. Calculus instructors can provide a useful perspective for future high school teachers (and other undergraduates as well) by giving more explicit attention to the way that general formulations about functions are used to express and reason about key ideas throughout calculus. Its central concepts, the derivative and the integral, are conceptually rich functions. Many formulas in calculus, such as the chain rule and integration by parts, are posed in general form rather than formulated separately for each function. The study of Taylor series shows how all differentiable functions can be approximated by polynomials.

The calculus reform movement has suggested differences in the way functions are treated in instruction. For example, there is a greater attention to multiple ways of looking at functions by (i) comparing algebraic, graphical, and numerical information about a function; and (ii) by highlighting their role in mathematical modeling. Experience in reform calculus courses, with their emphasis on conceptual depth, has the potential to be especially useful for future teachers.

Differential equations. A course in differential equations is required in some mathematics majors. Although often oriented towards physical science applications, this subject is also a good "practical" setting to strengthen undergraduates' understanding of functions, especially the role of functions in mathematical modeling. The whole notion of a differential equation is centered around an implicitly defined function that traces out a behavior governed by the differential equation. One sees solutions that are families of functions as in first-year calculus, but these families may have more complicated relationships than simply differing by a constant. Wronskians are connected with the notion of linearly independent functions. Laplace and Fourier transforms are linear operators on functions—defined in a way that may require a novel view of functions for many undergraduates. Power series solutions to differential equations introduce, in a very practical way, another useful way of representing functions.

Differential equations are now commonly studied with the assistance of a variety of graphic and computational software. Exposure to these new ways of working with functions will be valuable experience for future teachers.

Linear algebra. Although called linear algebra, this course can be taught with a substantial amount of linear analysis. It is important that undergraduates notice that functions play two different roles: they can be both linear transformations on vector spaces and elements of vector spaces. There are a variety of insights into functions that can be obtained by studying various vector spaces and linear transformations. For example, studying the vector space of power series and linear transformations, such as differentiation can help undergraduates appreciate the power of linear algebra in analysis.

Advanced calculus/introduction to analysis. An advanced calculus course or an elementary introduction to real analysis is an opportunity to give a rigorous foundation for future teaching about functions and calculus. Informal notions about Euclidean space, functions, and calculus that undergraduates have used for several years can be given sound formal definitions. The sort of syllabus that would be most valuable for prospective high school mathematics teachers would cover elementary topology of open and closed sets on the real line, sequences and series, properties of functions such as continuity, a formal treatment of the derivative and the Riemann integral, and basic properties of metric spaces. Key results like the Intermediate Value Theorem and the Mean Value Theorem can be proved in a more rigorous way than in a first-year calculus course.

Probability/statistics and discrete mathematics. Courses in these newer areas in the curriculum can and should make important and deep use of functions. It is easy to forget that central concepts in probability and statistics are closely tied to functions. For example, a random variable is a function on a sample space, and a statistic is a function of a set of random variables. In discrete mathematics courses, generating functions and recurrence relations (on integer-valued functions) present two new, intellectually rich aspects of the concept of a function. Making explicit connections between these specific examples and the function concept that occurs in other areas as well should help to provide mathematical insight that will be of use to future teachers and other undergraduates as well.

Functions in the capstone sequence. Although mathematics majors use functions in nearly every undergraduate course, it is recommended that prospective teachers to revisit the elementary functions of high school mathematics from an advanced standpoint in much the same way that they revisit algebraic and number system operations from a structural point of view. This sort of reflective look at functions and their unifying role in mathematics could be a prominent part of the capstone content course. The capstone study of functions can examine again the role of computer numeric and graphic tools in mathematical work—an important example of the connection between exploratory investigations and formal proof—and it can give undergraduates experience in the kind of complex problem solving required by authentic mathematical modeling.

Historical perspectives. Like many other mathematical ideas with broad applicability in modern mathematics, the function concept did not develop easily or early in the subject. The word "function" first appeared in a manuscript of Leibniz, and the first explicit definition of the term was by Johann Bernoulli who said, "One calls here a function of a variable a quantity composed in any manner whatever of this variable and of constants" (Siu, 1991, p. 105). However, many other distinguished mathematicians contributed to formulation of the concept, and it wasn't until late in the 19th century that the full generality and subtlety of the modern concept of function was explored. Study of this history will certainly give high school teachers appreciation for their students' difficulties in comprehending the concept that experienced mathematicians use almost without thinking. It also provides a case study in the way mathematical definitions evolve.

Common conceptions. Like early mathematicians, students very often come to believe that the only mathematical relations worthy of the name "function" are those that can be expressed by some sort of algebraic formula. They have difficulty imagining the variety of relationships that meet the formal definition of function, and few acquire the ability to think about functions as objects rather than processes for transforming inputs to outputs.

Recent research documents common cognitive difficulties in learning about and using functions (Dubinsky & Harel, 1992). The combination of mathematical and psychological issues involved provide yet another opportunity for mathematicians and mathematics educators to collaborate in capstone course activities that will help future teachers enter high school mathematics classrooms with deep personal understanding of this key concept and a sense of the instructional challenges it poses.

Applications. Undergraduate courses in calculus will give prospective teachers many encounters with ways that functions are useful in solving quantitative problems. They are less likely to provide a realistic experience in the mathematical modeling process that surrounds specific calculations. Since mathematical modeling has become a process that pervades problem solving and decision-making in the sciences, engineering, economics, and government, it is important for future high school teachers to have some experience with that kind of applied work.

Technology. Graphing calculators and computers are powerful tools in mathematical study and problem solving with functions, and the tools that are available become more impressive every year. Use of calculators and computers in precalculus mathematics and calculus is becoming very common in high school and collegiate mathematics courses. However, there is no consensus on the optimal way to introduce or use the technology. Activities in the capstone sequence for high school teachers should explore the variety of possible uses of calculators and computers in analysis—from numerical and graphic exploration and problem solving to formal symbolic operations in algebra, calculus, and linear algebra—and consider carefully the interplay of technology and formal reasoning methods.

Connections. Because the concept of a function has such general applicability in pure mathematics, its study offers opportunities to show the common mathematical structures in topics that are often perceived to be quite separate. For example, functions are often encountered first as models of linear, quadratic, exponential, or periodic variation—relating numerical domains and ranges. However, the same underlying concept provides a dynamic way of thinking about geometry in terms of transformations, and the transformation concept returns the favor by providing a way of connecting algebraic forms like x^2, ax^2, $(x+b)^2$, and $a(x+b)^2 + c$ and similar variations on e^x, $\ln x$, or $\sin x$. Functions also provide models for probability distributions and useful data transformations like log-log plots. A capstone experience that highlights those kinds of connections via functions will provide valuable insight to prospective teachers.

In traditional college preparatory curricula, the primary goal is preparing for study of calculus. Calculus is now commonly taught in advanced placement form at many U. S. high schools, so teachers should understand calculus well enough to make informed decisions about the content and emphasis of courses intended to prepare students for calculus and to be prepared themselves to teach high school calculus. Prospective high school mathematics teachers often enter undergraduate studies with a year of high school calculus. But even if they do all of their calculus at the collegiate level, they typically acquire only a mechanical and unsure grasp of the subject. Even those who tackle the challenge of advanced calculus or real analysis will acquire a fairly fragile basis for teaching calculus to high school students.

Like curricula and instruction in K–12 mathematics, collegiate calculus is evolving in response to emerging technology, diverse applications, and new ideas about teaching/learning environments. There is considerable hope that these reforms will lead to deeper student understanding and skill in the subject. Such an improvement would be very helpful for prospective high school teachers. Because the concept of a function is a primary building block for future study of calculus, it will be valuable for prospective high school teachers to examine fundamental calculus concepts like limits, continuity, differentiation, and integration with an eye to the conceptual difficulties of those subjects.

Data Analysis, Statistics, and Probability

Statistics has emerged recently as a core strand of precollege curricula. For many years middle and high school mathematics textbooks have included units on finite probability and a small amount of work on graphic display of data, with data summaries limited to the familiar mean, median, and mode. The situation is now very different. There is an Advanced Placement statistics exam for high school students. The American Statistical Association's Quantitative Literacy Project has developed data-driven curriculum modules that offer fresh approaches to high school topics in statistics, algebra, and geometry. The traditional emphasis on probability-based statistical inference has given way to introductory statistics material that focuses on using data to "gain insight into real problems" (Moore & McCabe, 1999, p. xvii). This new conception of the content of probability and statistics is accompanied by commitment to teaching the subject in ways that reflect the practice of statistics. To be prepared to teach in this way, prospective teachers should have experience formulating questions, devising data collection protocols, and analyzing real data sets that result from their own investigations or from the data collection of others.

Readers are urged not to view the introduction of statistics topics as a watering down of the school mathematics curriculum. Probability and statistics are mathematically rich subjects that can be taught in a fashion that strengthens students' understanding of algebra and functions. Working with the concepts of the probability of an event, a random variable, and a statistic all involve solid mathematical reasoning and considerable thoughtfulness. Serious high school study in this area, including in a senior semester-long course for students headed for social science majors, has much to recommend it. Although data analysis has its simple descriptive side, such as constructing stem-and-leaf plots and calculating summary statistics, it also involves conceptually rich issues. There are deep and subtle questions in exploration of data distributions, modeling relationships among two or more variables, designing complex surveys or experiments to reduce bias and variability, and inferential reasoning (hypothesis testing and estimation through confidence intervals).

Curricula for the mathematical preparation of high school teachers must include courses and experiences that help them appreciate and understand the major themes of statistical practice:

- Exploring data: using a variety of standard techniques for organizing and displaying data in order to detect patterns and departures from patterns.

- Planning a study: using surveys to estimate population characteristics and designing experiments to test conjectured relationships among variables.

- Anticipating patterns: using theory and simulations to study probability distributions and apply them as models of real phenomena.

- Statistical inference: using probability models to draw conclusions from data and measure the uncertainty of those conclusions.

Probability has important applications outside of statistics. Thus, prospective high school teachers should also:

- Understand basic concepts of probability such as conditional probability and independence, and develop skill in calculating probabilities associated with those concepts.

Statistics is now widely acknowledged to be an extremely valuable set of tools for problem-solving and decision-making. But, despite the production of interesting statistics materials for schools, it has been hard to find room for the subject in curricula dominated by preparation for calculus.

Statistics and probability courses. Future high school mathematics teachers need at least two courses in probability and statistics. The most natural pair would be a calculus-based survey course in probability and statistics and a course in data analysis.

Survey of probability and statistics. The survey course is typically aimed at a broad audience of majors in mathematics, computer science and engineering.

Guidelines for such a course in the 1981 CUPM Recommendations for a General
Mathematical Sciences program suggest starting with a survey of elements of data
analysis. This survey might be reduced in light of the proposed separate data
analysis course. Then there is an introduction to probability—sample spaces, independence, conditional probability and Bayes' Theorem, common discrete and
continuous probability distributions, and the Central Limit Theorem—followed by
an introduction to mathematical statistics—common sampling distributions, point
estimation, tests of hypotheses, and confidence intervals. Note that this course
should spend only a modest amount of time on combinatorial probability problems, because combinatorial enumeration will be treated in the discrete mathematics course. Modern versions of this course use technology to demonstrate properties
of probability distributions and inferential procedures.

Data production and analysis. This course should include elements of design
of experiments and sample surveys, parametric inference (typically assuming a
normal distribution of the population) on population means and proportions, the
chi-square test for goodness-of-fit, association and homogeneity of proportions, regression analysis, and analysis of variance. If time permits, modern techniques
for categorical data analysis (such as logistic regression) and some non-parametric
procedures (such as the Wilcoxon rank sum test) might be introduced. Modern
computer-intensive techniques such as the bootstrap could be introduced in such a
course. The course should require a major data project from each undergraduate
or from groups of undergraduates.

Probability and statistics in the capstone sequence. In addition to the
two courses suggested above, a segment of the capstone sequence for teachers should
address issues in data analysis, statistics, and probability. As is the case for algebra,
geometry, and analysis, the capstone work could address history, common student
conceptions, applications, technology, and connections.

Historical perspectives. In comparison with algebra and geometry, the history
of probability and statistics is relatively short. The modern theory of probability
began with the work of Fermat and Pascal in the mid-17th century and it was
not until the 1930s that Kolmogoroff laid axiomatic foundations for the theory.
Statistics also traces its roots to events in the 17th century, but formal inference
is a 20th-century development and data analysis has been almost reinvented in the
last several decades, as computers provided new tools for collection, summary, and
representation of distributions. Some knowledge of this history of probability and
statistics is a very useful resource for high school teachers.

Common conceptions. In contrast with other branches of mathematics which
tend to focus on deterministic relationships, probability and statistics provide tools
for reasoning about chance and variation. Research has revealed an interesting array
of cognitive difficulties that most people have in dealing with this uncertainty. For
example, it is well known that most people will judge that the outcome "HTTHHT"
is more likely than the outcome "HHHTTT" in six tosses of a fair coin. Most people
will judge that the outcomes "at least 4 heads in 6 tosses of a fair coin" and "at least
400 heads in 600 tosses of the same coin" are equally likely (Shaughnessy, 1992).
Few people can accurately explain the meaning of confidence intervals or statistical
significance, even after introductory courses in probability and statistics. Preparing
high school mathematics teachers for their responsibilities in teaching statistics

and probability will require attention to both the mathematical and psychological aspects of these and many other common conceptions about the subject.

Applications. Current precollege teaching of statistics is generally based in analysis of data that arise from measurements in real or easily imagined situations. Statistical reasoning and arguments appear almost daily in news media and are available in classroom-usable form at a number of Web sites (e.g., `http://www.dartmouth.edu/~chance/chance_news/news.html`). Thus most future high school teachers won't need much help in connecting statistics and probability concepts to realistic teaching materials. At the same time, they probably do need help in conceptualizing the relationship between data and mathematical models that is at the heart of statistical practice (Moore, 1990, p. 99). This statistical modeling activity can serve to reinforce important concepts in algebra and functions.

Technology. Calculators and computers are now a fundamental part of doing and learning about probability and statistics. They have revolutionized the practice of statistics in the ways that they allow analysis and representation of huge data sets. They have transformed statistics education in much the same way. In preparing to teach probability and statistics to high school students, prospective teachers need to become proficient in the use of standard software tools for simulation, data analysis, data modeling, and inference. They also need to learn the common difficulties that arise when those tools are misused or trusted without careful consideration of the techniques being applied to a given problem situation. Savvy technology use will almost certainly not develop after one or even two beginning statistics courses, but it is a natural consideration in a capstone course for teachers.

Connections. The preceding discussions of algebra, geometry, and functions have indicated a number of ways that probability and statistics connect to those core mathematical topics. The most powerful of those connections is due to the fundamental role of data (which Moore describes as "numbers with context") in statistical practice. Algebra and functions provide models for data patterns; interesting data sets provide settings for thinking about classes of functions and their symbolic representations. The connections in both directions are very effective means of engaging high school students in significant mathematical study. Teachers who understand and appreciate these connections will have a very useful teaching resource.

Discrete Mathematics and Computer Science

The increasing mathematization of disciplines outside of the physical and engineering sciences has included some impressive new applications of classical mathematics. More significantly, it has also stimulated development and use of a number of new areas that are often collectively referred to as discrete mathematics. Graph theory, enumerative combinatorics, finite difference equations, iteration, and recursion are core topics in this area. They play fundamental roles in computer science, operations research, economics, and biology. Ideas from the theory of games, fair division, and voting are in the part of discrete mathematics that provides tools for social decision-making (Kenney & Hirsch, 1991).

Many of the central ideas of discrete mathematics are as accessible to high school students as parts of classical curricula, and many students find the applications of discrete mathematics very engaging. Furthermore, discrete mathematics topics have important connections to other core strands of the high school curriculum. Recurrence relations provide a powerful strategy for representing and reasoning about functions, graph theory extends visual thinking and geometric reasoning, combinatorics has obvious applications in probability, and game theory/conflict resolution problems make use of matrices.

In light of the emerging importance of topics and methods in discrete mathematics, prospective high school teachers need a course that introduces them to the mathematical ideas, methods, and applications of topics in the following areas:

- Graph theory.

- Enumerative combinatorics.

- Finite difference equations, iteration, and recursion.

- Models for social decision-making.

An upper-division course addressing these topics would also serve majors in mathematics, computer science, and engineering. Except for the last topic, this proposed course is similar to the Discrete Methods course syllabus in the 1981 CUPM Recommendations for a General Mathematical Sciences Program. However, the proposed syllabus still represents a broad agenda, and instructors in the course will have to make choices of topics to survey and topics to develop in depth.

Given the large number of high school students who now enter careers that involve computing and the pervasive use of discrete mathematics in the design of computer hardware, algorithms, and software, it is incumbent on the mathematics community to give greater attention to the mathematical foundations of computer science and computer programming in the schools and in the preparation of future teachers. For prospective high school teachers, a solid one-semester introduction to computer science is essential for connecting mathematics to our computer-driven world.

Most prospective high school mathematics teachers enter college with rudimentary knowledge of computer programming and little knowledge of computer science. Teacher preparation programs commonly include a computer science requirement, often a programming course. Future high school mathematics teachers should have an introduction to computer science that addresses three themes:

- Discrete structures (sets, logic, relations, and functions) and their applications in design of data structures and programming.

- Design and analysis of algorithms, including use of recursion and combinatorics.

- Use of programming to solve problems.

Mathematical Thinking

The preceding descriptions of desirable content preparation for high school mathematics teachers comprise an ambitious agenda, extending and modifying the focus of typical current programs. Most of the mathematics outlined above appears now in one or another of the standard undergraduate courses. Each strand description suggested some new perspectives to be taken in presentation of that content so that students could use advanced mathematical study to gain deeper insight into the structure and uses of high school mathematics topics. But teachers of mathematics convey more than the technical skills of the subject.

By their choice of classroom mathematical tasks and activities and by their personal attitudes and behavior as mathematicians, teachers shape their students' perceptions of what it means to reason mathematically. They suggest and demonstrate the habits of mind that are characteristic of a mathematical approach to questions (Cuoco, Goldenberg, & Mark, 1998). Undergraduate experiences in mathematics should help future teachers understand and practice broadly applicable habits of mathematical thinking: (i) asking and exploring interesting mathematical questions; (ii) framing mathematical concepts and relationships in clear language and notation; (iii) constructing and analyzing proofs; (iv) applying mathematical principles in other disciplines.

In his introduction to the book *On the Shoulders of Giants: New Approaches to Numeracy*, Lynn Steen (1990) remarked that,

> Mathematics is not just about number and shape but about pattern and order of all sorts. Number and shape—arithmetic and geometry—are but two of many media in which mathematicians work. Active mathematicians seek patterns wherever they arise. (p. 2)

In their search for and application of patterns, mathematicians look for attributes like linearity, periodicity, continuity, randomness, or symmetry. They take actions like representing, experimenting, modeling, classifying, visualizing, computing, and proving. They use abstractions like symbols, infinity, logic, similarity, recursion, and optimization. And they look for contrasts like discrete versus continuous, finite versus infinite, algorithmic versus existential, exact versus approximate, or stochastic versus deterministic patterns. In urging consideration of curricula organized around habits of mind, as much as lists of topics, Cuoco, Goldenberg, and Mark (1998) suggested some similar perspectives on the nature of mathematical work.

In rethinking the mathematical education of future high school mathematics teachers, it seems important to consider also such an agenda of habits-of-mind goals. Most mathematics faculty probably agree with such objectives and even argue that their courses include remarks or assignments designed to cultivate these desirable habits and dispositions. However, students often emerge from their undergraduate experiences with, at best, an unarticulated sense of what it means to be a mathematician. More explicit attention to this aspect of mathematical education may be needed in teacher preparation coursework.

Many mathematics programs also have a sophomore-level course with a title like Foundations of Higher Mathematics that focuses explicitly on developing mathematical thinking, especially skill in reading and writing proofs. These courses were popular in the 1970s as the level of theory in the mathematics major was being raised. Then in the 1980s, many of the courses were dropped as attention focused on adding breadth in the mathematical sciences. In the past 15 years, in response to serious deficiencies in mathematical thinking among students in junior-level courses such as Abstract Algebra and Introduction to Real Analysis, many departments have reinstated or started such courses. The decision by so many mathematics departments to offer such a course is a strong argument in favor of its value. There are now several well-received texts for this course, although some faculty like to fashion their own individual course from selected readings. Some of these courses and texts focus on mathematical foundations in set theory, logic, relations, and number systems. Others cover number systems and some foundations material and then move on to selected problems in linear algebra, analysis and geometry.

Historical and Cultural Perspectives

Because of its enormous practical value, mathematics is frequently taught as a collection of technical skills that are applicable to specific tasks and often presented without reference to the intellectual struggles that led to contemporary understanding. The few personal names attached to particular mathematical results tend to suggest that mathematics is largely a product of European men.

Over the past several decades there have been important historical and cultural studies that reveal a rich and culturally diverse history of mathematics (Katz, 1992; Trentacosta & Kenney, 1997). We now know a great deal about the difficulty that very good mathematicians had with apparently simple ideas like negative or complex numbers, limits and continuity, and symbolic notation for algebra. We also realize that many non-European cultures have made sophisticated and significant contributions to mathematics.

Few prospective high school teachers come to undergraduate studies with much, if any, understanding of these issues. Some mathematics departments have the faculty expertise and undergraduate enrollment to offer history of mathematics courses. Some may choose to build historical background into the proposed capstone course. In either case, future high school teachers will be well-served by deeper knowledge of the historical and cultural roots of mathematical ideas and practices.

Experiences in Learning Mathematics

Typical collegiate mathematics instruction is associated with some implicit assumptions about learning. One should first hear a knowledgeable person explain the what and how and why of some new idea, then go off alone to see if one understood enough of the lecture to apply ideas and skills appropriately and effectively to tasks similar to those in the teacher's examples. The interaction between teacher and student in the typical college classroom is often limited to the silent thoughts of the students watching and listening. Chalk and talk are the primary instructional media. Emerging from extended experiences with this style of mathematics teaching and learning, prospective high school teachers frequently enter their methods courses and field experiences with with two main models of teaching: one from college and one from precollege schooling. Often, neither of these models suffices for the demands of current high school curricula.

The insight and personal enthusiasm of an inspiring lecturer often contribute to successful experiences for college students. However, in many cases the results are far from optimal. When even the most lucid explanations don't effectively communicate the big ideas of a mathematical topic, students are often reduced to memorization of problem-solving routines through imitation of worked examples in lectures and text materials.

Over the past 10 to 15 years, some instructors have experimented with ways to have undergraduates more actively involved in the intellectual work of class meetings. In such classes, students collaborate with each other on investigations that pose challenging problems with solutions that reveal important mathematical principles. They use calculators and computers as tools in those investigations and in composing reports of their findings. Such changes in teaching, if appropriately made, can result in better learning—more conceptual understanding, improved abilities to communicate orally and in writing, and to solve problems (see, e.g., Darken, Wynegar, & Kuhn, 2000; Harel & Sowder, 1998; Schoenfeld, 1985; Springer, Stanne, & Donovan, 1999). Moreover, such instruction gives prospective teachers experience with the kind of instruction they will be asked to use in their future careers.

Conclusion

The recommendations in this chapter offer a curriculum for the mathematical preparation of teachers that, compared to the chapters on elementary and middle school mathematics teachers, does not call for radical changes from past recommendations by groups such as the MAA Committee on the Mathematical Education of Teachers. The main new component recommended here is a year-long capstone course for prospective high school teachers that connects the major strands in courses for mathematics majors with high school mathematics. Other recommendations are: some refocusing of upper-division courses in abstract algebra and geometry, a new course in statistics, and some enhancement of lower-division courses.

Many of these recommendations may be as appropriate for all mathematics majors as they are for prospective teachers. However, the capstone sequence is aimed specially at future teachers. This course is an opportunity to look deeply at fundamental ideas of mathematics, to connect topics which students often see as unrelated, and to develop the important mathematical habits of mind. It connects logically to the coursework in mathematics education that also makes a strong positive contribution to the mathematical understanding and pedagogical skills of teachers (Monk, 1994). The design and teaching of this course is a natural point for collaboration of mathematics and mathematics education faculty who have joint responsibility for teacher preparation.

References

Bransford, J., Brown, A., & Cocking, R. (Eds.). (1999). *How people learn: Brain, mind, experience, and school.* Washington, DC: National Academy Press.

Begle, E. (1979). *Critical variables in mathematics education: Findings from a survey of empirical literature.* Washington, DC: Mathematical Association of America.

Boyer, C. B. (1991). *A history of mathematics* (revised by U. C. Merzbach). New York: Wiley.

Confrey, J. (1990). A review of the research on student conceptions in mathematics, science, and programming. In C. B. Cazden (Ed.), *Review of research in education, 16,* 3-56.

Cuoco, A., Goldenberg, E. P., & Mark, J. (1996). Habits of mind: An organizing principle for a mathematics curriculum. *Journal of Mathematical Behavior, 15*(4), 375-402.

Darken, B., Wynegar, R., & Kuhn, S. (2000). Evaluating calculus reform: A review and a longitudinal study. In E. Dubinsky, A. H. Schoenfeld, & J. Kaput (Eds.), *Research in collegiate mathematics education IV* (pp. 16-41). Providence, RI: American Mathematical Society.

Darling-Hammond, L., Wise, A., & Klein, S. (1995). *A license to teach: Building a profession for the 21st century schools.* Boulder, CO: Westview Press.

Dubinsky, E. & Harel, G. (Eds.). (1992). *The concept of function: Aspects of epistemology and pedagogy* (MAA notes no. 25). Washington, DC: Mathematical Association of America.

Harel, G. & Sowder, L. (1998). Students' proof schemes: Results from exploratory studies. In A. H. Schoenfeld, J. Kaput, & E. Dubinsky (Eds.), *Research in collegiate mathematics education III* (pp. 234-283). Providence, RI: American Mathematical Society.

Heid, M. K. (1997). The technological revolution and the reform of school mathematics. *American Journal of Education, 106,* 5-61.

Henderson, D. (2000). *Experiencing geometry: In Euclidean, spherical, and hyperbolic spaces.* Englewood Cliffs, NJ: Prentice-Hall.

Katz, V. (1993). *A history of mathematics: An introduction.* New York: Harper Collins.

Katz, V. (Ed.). (2000). *Using history to teach mathematics: An international perspective.* Washington, DC: Mathematical Association of America.

Kenney, M. J. & Hirsch, C. R. (Eds.). (1991). *Discrete mathematics across the curriculum, K-12.* Reston, VA: National Council of Teachers of Mathematics.

Kieran, C. (1992). In D. A. Grouws (Ed.), *Handbook of research on mathematics teaching and learning* (pp. 390–419). New York: Macmillan Publishing.

Ma, L. (1999). *Knowing and teaching elementary mathematics: Teachers' understanding of fundamental mathematics in China and the United States.* Mahwah, NJ: Erlbaum.

Mathematical Association of America. (1989). CUPM Recommendations for a General Mathematical Sciences Program. In *Reshaping college mathematics* (MAA notes no. 13). Washington, DC: Mathematical Association of America. (Original work published 1981)

Mathematical Association of America. (1983). *Recommendations on the mathematical preparation of teachers.* Washington, DC: Author.

Monk, D. A. (1994). Subject area preparation of secondary mathematics and science teachers and student achievement. *Economics of Education Review, 13*(2), 125-145.

Moore, D. S. (1990). Uncertainty. In L. A. Steen (Ed.), *On the shoulders of giants: New approaches to numeracy* (pp. 95-138). Washington, DC: National Academy Press.

Moore, D. S. & McCabe, G. (1999). *Introduction to the practice of statistics.* New York: Freeman.

National Center for Research on Teacher Education. (1991). *Final report.* East Lansing, MI: Author.

Schoenfeld, A. H. (1985). *Mathematical problem solving.* Orlando, FL: Academic Press.

Shaughnessy, M. (1992). Research in probability and statistics: Reflections and directions. In D. A. Grouws (Ed.), *Handbook of research on mathematics teaching and learning* (pp. 465-494). New York: Macmillan Publishing.

Shulman, L. S. (1986). Those who understand: Knowledge growth in teaching. *Educational Researcher, 15*(2), 4-14.

Siu, M. K. (1991). Concept of function—its history and teaching. In F. Swetz et al. (Eds.), *Learn from the masters* (pp. 105-121). Washington, DC: Mathematical Association of America.

Springer, L., Stanne, M. E., & Donovan, S. S. (1999). Effects of small-group learning on undergraduates in science, mathematics, engineering, and technology: A meta-analysis. *Review of Educational Research, 69*(1), 21-51.

Steen, L. A. (1990). Pattern. In L. A. Steen (Ed.), *On the shoulders of giants: New approaches to numeracy* (pp. 1-10). Washington, DC: National Academy Press.

Swetz, F., Fauvel, J., Bekken, O., Johansson, B., & Katz, V. (Eds.). (1995). *Learn from the masters.* Washington, DC: Mathematical Association of America.

Trentacosta, J. & Kenney, M. J. (Eds.). (1997). *Multicultural and gender equity in the mathematics classroom: The gift of diversity.* Reston, VA: National Council of Teachers of Mathematics.